HOW DOGS MAKE US FALL IN

LOVE WITH THEM

AND ELEVEN OTHER SHORT STORIES

FROM THE FRONTIERS OF BIOSCIENCE: 2015

JOHN R. SPEAKMAN

About the author and the articles

UK born, John R. Speakman is a 1000 talents 'A' Professor at the Institute of Genetics and Developmental Biology, Chinese Academy of Sciences in Beijing, China, and a Professor in the Institute of Biological and Environmental Sciences at the University of Aberdeen, in the UK. He has lived in Beijing, China since 2011. In 2013 he also became a features writer for the Chinese monthly popular science magazine 'Newton'. The articles in this collection were written in English and were previously translated into Chinese and appeared each month in 'Newton'. This book is a compilation of the original English versions of the stories, modified for a western audience. The 'further reading' section at the back of the book lists the original sources on which the articles were based.

By the same author

Pandas – dead end or dead wrong? and eleven other short stories from the frontiers of bioscience 2014.

For Mary

Acknowledgements

I am grateful to Lina Zhang for translating these articles into Chinese and to Newton magazine for publishing the Chinese versions, while allowing me to retain copyright on the English versions of the articles. Thanks to Kristina Campbell for the photo of her dog Gus, and Teresa Valencak for the photo of her dog Poppy.

Cover photos:	**Front**	**Kristina Campbell.**
	Back	**Catherine Hambly.**
Photos inside:	**Body images**	**Mark Faries**
	Panda	**John Speakman**

Contents

Beyond comprehension

January 2015

Imagine for a moment that you are sitting at home eating a piece of fruit. You put a slice of the fruit down on the table in front of you, and a small fruit fly comes along and lands on it. Consider now a problem. What if I had asked you to teach the fly some simple maths. Like how to add together any two numbers that when combined sum to less than 10. So you might say to the fly (in some manner it can understand) what is three plus four, and you would expect it to answer (through a suitable translation device) seven. Or you might ask it two plus six, and expect the answer eight. Such simple addition is a rudimentary problem that almost all humans above the age of ten can solve. Yet for a fly this is a problem of monumental complexity. Quite possibly, even if we could rig up some way of translating the problem into a 'language' that the fly could understand, and we could build a device to translate the answer, this is a problem that a fruit fly is simply incapable of solving, because it does not have the brain capacity to do so. No matter how long you spent trying, you might never achieve it. Yet importantly this (or at least a very similar problem) is a real issue that we may face in twenty to thirty years time, if things progress in the way they have been doing for the past few decades.

By way of explanation let me take you on a little detour. At the end of October last year I had the pleasure to attend a talk by one of the most successful scientists of the modern era: Jun Wang, a co-director of the sequencing company BGI (short for Beijing

Genomics Institute but now located after a spell in Hangzhou in Shenzen). Jun Wang has won many accolades for his work, and his CV boasts a collection of over 100 publications in the top journals *Nature* and *Science* – including being featured more than 15x on the front covers of these journals. He was recently named as one of the top 10 scientists in the world aged under 40. In his interview with *Nature* when awarded this honor, he made a quote that he also used at the start of his speech. "Thank you for your introduction" he said "...but I dont have a brain. Only muscle." The audience laughed, rather nervously, and I saw several people exchanging confused glances. What on earth could he mean? Wang went on to talk about the cost of doing DNA sequencing work. The human genome project started with the goal of sequencing an entire human genome in the early 1990s. BGI was actually set up in 1999 to participate in the human genome project, as the Chinese representative, and it received funds from the US government to take part. In total the Human genome project involved an effort that took over 10 years to complete (a draft was published in 2000 and the final sequence in 2003), involved many thousands of scientists and in aggregate is estimated to have cost about 3.8 billion US dollars. The result was a sequence of pooled DNA from 5 individuals , but predominantly from a single anonymous donor, code named RP11, who came from Buffalo, New York State in the USA. A rival project run by the private company, Celera, which was only established in 1998, used a different sequencing approach, and came up with a draft sequence at the same time as the publically funded human genome project, at a cost that was about 300 million US dollars, this time sequencing an aggregate sample of DNA pooled from 20 individuals, including Celera's CEO Craig Venter. The first published sequence, unambiguously from a single individual, was published in 2007 and concerned the genome of James Watson, one of the original discoverers of the structure of DNA. Driven by advances in technology, the cost of generating Watson's genome in 2007 had already fallen to around 10 million dollars. This was a decline by a factor of >10 in just 5 years, compared to the Celera cost.

In 2007 the company Applied Biosystems introduced a new technology which changed the face, and price, of sequencing. Since 2008, according to figures produced by the National Human Genome Research Institute the cost of sequencing has fallen by about an order of magnitude every year. Hence in 2009 it cost about 1,000,000 USD (680 wan RMB), in 2010 when Steve Jobs of Apple had his sequence done, it was 100,000 USD, in 2012 it cost 10,000 USD and in February last year it was announced by the company Illumina that it would be possible using their Hi-seq X10 system to get your full sequence done for the magic 1,000 USD figure. Hence in just over a decade since the human genome and Celera sequencing projects were published in *Nature* and *Science*, we have gone from an exercise costing a large fraction of the US science budget and requiring a global collaboration of thousands, to something that was affordable only to the super-rich, to something that currently falls within reach of being an upmarket birthday present for large swathes of the population. If you are thinking of a novel gift for your partners birthday next time this could be it..but read on, you may want to wait a bit!

This precipitous decline in the cost of sequencing has been suggested to be something analogous to the technological developments that have happened in the electronics world, generally called Moore's law. In fact this year sees the 50th anniversary of the publication of the article that Moore's law originated from. In 1965, Moore was the director of research and development at Fairchild semiconductor, which ultimately became Intel. In his article, Moore speculated on the future development of the computer industry. He observed that there had been an exponential relationship between device complexity and time. In this case he observed that roughly every 2 years the density of components on a microchip was doubling. In 1965 the densest microchip contained just over 2000 components. Moore speculated that by 1975 this would mean that a chip might contain 65,000 components. Although Moore originally speculated that the exponential growth might continue for only 10 years or so, the 'law' has stood the test of time for 50 years, and the numbers of transistors now on chips is counted in millions, and this continues to double on a 2-3 year average time frame. Predictions of how long Moore's law will continue to pertain into the future seem to always be around 10 years!

So, back to sequencing costs. In his talk Jun Wang suggested that there was no reason why the cost of sequencing should not continue to fall, even if the rate at which it falls is significantly slower. That would mean, if the rate halved, by 2017 we will perhaps have the 100 US dollar genome, by 2019 the 10 US dollar genome and by 2021, almost exactly twenty years on from the draft human genome paper, it may just be one dollar to get your entire sequence done. At which point Wang suggested it would actually be effectively free....but what you would need would be thousands of sequencing machines to do the work and that is what he meant by 'muscle'. BGI currently has more sequencing capacity than any other organisation. It has more muscle.

Wang predicted that the costs of sequencing will decline to the point where the cost of doing it is far less than the value of the information it yields. Everyone, he confidently predicted, should be sequenced, not because there is a reason to do it, but simply because it is so cheap to do it and collect the data, and the data will be valuable in its own right. And therein lies the key to the other part of Wang's statement. I have no brain..only muscle, because this approach is a quiet revolution to the whole way that we do science. Between about four-hundred and one hundred years ago science was a voyage of discovery – often literally. Think of the voyage of the Beagle, for example. Scientists were data collectors. They collected information about the world because at the time we actually knew so little that we needed the raw data on which people could start to build theories about the way the world worked. The theory of evolution (approximately 150 years ago), Newton's laws of mechanics (300 years ago), Einstein's theory of relativity (100 years ago) and the periodic table of elements (Mendeleev 150 years ago) all emerged from this period of basic observation and speculation about the way things worked. At the same time, of course, there were lots of other ideas around that were credible ideas, but they were incorrect. We forget them now because they have been discredited and are no longer taught, but they included for example the ideas of alchemy and turning metals into gold, the inheritance of acquired characteristics, and the phlogiston theory of chemistry. How did we reject these ideas, and settle on the ones we are now familiar with? It was because of the greatest invention of that age – and some suggest the greatest invention of mankind ever, and that was the scientific method. The use of theory to generate hypotheses, and then the derivation of experiments to test these hypotheses to provide information that confirms or rejects the original idea. In the 1980s when I was a young scientist it was impossible to get even a small 500 dollar travel grant without first having a clear hypothesis. But the human genome project stepped outside this framework and started a revolution. That is because the human genome project and the Celera project together involved the spend of over 4 billion dollars, but neither of them had a hypothesis. It was done in the spirit of the trip to the moon - simply for the challenge of finding out. And that is the revolution that Jun Wang continues today with his current ambition to sequence ten thousand genomes, and then a million, and eventually as it becomes virtually free - everyone. There is no hypothesis. There is no brain...only muscle. The goal is only to generate information. The reason it is argued that we need such a revolution is because our understanding of the way that this system works is so rudimentary that we are like the early explorers. They travelled the globe to find new lands and wondrous animals. They had no hypotheses, just meticulous observational skills. Now, we travel the genome in a similar adventure, generating information that we will ultimately be able to coalesce into theories and hypotheses that will revolutionise our understanding of the way life works – and what happens when it goes wrong (like when cells become cancerous, or when individuals develop heart disease or diabetes)...and the biggest question of all, do we need to age and die?

Or will it? It is a romantic vision of the future. But there is a problem. Already we can generate sequence data for a person's genome, the details of the RNA that is transcribed from the DNA (called their transcriptome), the translated proteins this generates (the proteome) and the resultant metabolic changes in chemicals and lipids throughout the body (the metabolome and the lipidome) respectively. These are the so-called –omics technologies and the data that is produced comes in staggering quantities. A genome is 3 billion letters of code. The transcripome may involve millions of reads of RNA transcribed from 1000's of genes – and this is dwarfed by the complexity of the proteome and the myriad of post production modifications. Just presenting this information is a challenge. Many scientific papers already show networks of genes and metabolites interacting in a massive web that looks nice and very complex, but actually doesn't really tell us anything much about what is happening – apart from that it is very complex. One paper I recently reviewed at a top journal had 184 pages of gene lists provided as supplementary material. This large information flow may already be at a point where we already have too much information for us to process and fully understand – or even partially understand. More muscle may sound good, but it is probably not what is actually needed. Fortunately there may be a solution – but it comes with a bitter twist. Remember Moore's law in its original formulation regarding the number of transistors on a chip, and hence the processing power of computer chips. Already last year a computer was built that has the same number of connections as the human brain – but it cost 40,000 US dollars to build it. In ten years time it is projected that such a computer will cost only about 1000 dollars. Think about it, in 2025, your 'smart' phone will literally be as smart as you are – for some people their phones will be considerably smarter! If the trend continues, ten years after that the standard desktop computer will have the intellectual capacity of the entire combined 7 billion members of the human race! And maybe that's what we will need to provide the brain, to complement Jun Wang's muscle, and thereby make sense of all the data that will be produced when genome sequencing costs just a dollar. The problem will be that if we give the task to an intelligent machine with the combined intellectual capability of the whole of mankind, to use all the data that is being generated from the -omics technologies, to resolve how the human system works, and therefore how we can prevent things going wrong, the super-intelligent computer may not be able to communicate the answer to us. We may just be too stupid to understand it. Remember the fly on the fruit that you were trying to teach 3 + 4 = 7. In 20 years time that fly may be you, looking up at a super-intelligent computer that is trying to tell us how to solve the problem of aging, but it is just completely beyond our comprehension.

UPDATE

In July 2015 Jun Wang stepped down from his position as CEO of BGI to set up an artificial intelligence capable of analysing 100 million genomes with the goal of working out how to make people live healthier and longer.

The lottery you hope you will never win

February 2015

It took my friend Jim about three years to die. When he first told me about it, his problem was already a year along the road. It was a rollercoaster ride. At that stage, one year into his journey, they had already been pumping his leg full of chemicals to try and kill it. It looked like this course of action had been a success. When he told me, he spoke about it like it was something in the past. Something he had survived. We drank a bottle of wine together to celebrate. But then six months later they discovered some lumps in his groin. He needed more therapy. This time the drugs made him feel sick. But once again it seemed he had beaten it. Then about a year later he started to get tired at his work. He found it hard to make decisions on even the most simple things, and finally one evening he came home from work and started to cook a meal, but just froze. It took him 30 minutes to decide to boil the kettle. He went to hospital. It was back. But this time in his brain: where the drugs couldn't reach it. The only option was surgery. As he waited for the surgery in hospital we sent him a 'get well soon' card. But what do you write to someone who has terminal brain cancer? 'Get well soon' seems a bit too simplistic. My colleague and I decided on something unorthodox. Before he got ill we had been working with Jim on a scientific paper. Because of his illness we had stalled on writing it up. "Come on Jim" we wrote. "This is really the most pathetic excuse to get out of writing up that paper". It was a gamble, but later after the surgery I spoke with him and he told me that the card we sent him was one of the few things in those bleak days that had made him laugh. There wasn't much to laugh about. Three months later he was dead.

In the two minutes that it took you to read about my friend Jim's death, approximately another 40 people have died of cancer. In fact during 2012, the latest year for which we have complete data from across the globe, 8.2 million people died of it, which works out at one person on average every 3.8 seconds. In the modern world, cancer as a killer is exceeded only by heart disease. Most people appreciate that the causes of cancer are complex, but a model that has been around for some considerable time is the idea that it is caused by a combination of genetic and environmental factors. Exposure to certain environmental factors greatly increases the risk of developing certain types of cancer. The most obvious link is between smoking and lung cancer. People who smoke 20 cigarettes per day have about a 25x greater risk of developing lung cancer than people who have never smoked. But not everyone who smokes develops lung cancer. There is no simple one to one correlation. This fact enabled the tobacco companies to claim that smoking didn't cause cancer for decades. The problem was that in the 1950s and 1960s the major killers that had been tackled by medical establishment were infectious diseases. Infectious diseases are characterised by a one to one relationship with mortality. If you get ebola and you die, it is pretty clear that you died of ebola. It only takes a week or so to kill you. People who don't get ebola, don't die of ebola. But cancer is different. If you smoke you don't instantly develop lung cancer and die a week later. It takes years, or even decades for the lung cancer to develop. Plus some people who smoke heavily don't get it at all. The oldest recorded person ever, a French woman called Jeanne Calment, who died aged 122 years and 164 days in 1997, had smoked from the age of 21 until she was 117 years old. The person who took over from Calment as the oldest living person when she died, and who is currently the fourth eldest person ever, Marie-Louise Meilleur, smoked into her 90s. Plus on the other hand others who don't smoke get lung cancer anyway. Ten percent of people who develop lung cancer do not smoke. Who could tell therefore whether a person who smoked and developed lung cancer 20 years later wouldn't have developed lung cancer anyway, even if they hadn't started smoking. The causality if fuzzy. The supposed reason for this fuzziness is that we all carry genetic variations that make us more or less likely to develop certain cancers. It is suggested that it is the combination of these genetic and environmental exposures that determines whether cancer will develop or not. This has been likened to the risk of having a car accident. The risk you will have a car accident depends in part on the road conditions (an environmental factor). When it is icy your risk of an accident is increased, but it isn't inevitable that you will crash. Another factor however is the quality of your car. If the brakes are defective then you are more likely to crash whatever the road conditions, but again it isn't inevitable. Defective brakes are like a bad genetic mutation. Drive in bad road conditions with defective brakes and the risk of an accident is massively increased. It is the gene by environment interaction that is important.

One of the big promises of the human genome project was the idea of genetically personalised or precision medicine. That is we could tailor medical treatments to a persons individual genetic make-up. If you knew, for example, that you had a genetic mutation that would make you more susceptible to lung cancer if you smoked, then your medical practitioner could warn you in advance that smoking was a very bad lifestyle choice for you. To an extent this era is already with us. Perhaps the most celebrated example has been the real life medical adventure of the American actress Angelina Jolie. In 2013 Jolie announced that she had had a double mastectomy (removal of both of her breasts). The reason for this dramatic surgical intervention was not because Jolie had anything wrong with her breasts. It was done as a precaution because she has a genetic mutation in a gene called BRAC1 which gave her an estimated 87% chance of developing breast cancer over the rest of her life. She inherited this mutation from her mother. BRAC1 (and BRAC2) are genes that encode proteins that assist in the repair of damage to DNA. When they are defective the repair function is impaired and the result is an increased risk of cancer. The risk is increased for several different cancers but for some reason this mutation greatly increases the risk of breast and ovarian cancer. In fact Angelina Jolie's mother had breast cancer and died of ovarian cancer at the age of 56 in 2007, her aunt died of breast cancer in 2013, and her grandmother also died of ovarian cancer. By removing the tissue that is susceptible, Jolie was able to dramatically reduce the risk of developing cancer in that tissue to less than 5%. This was only possible because of our understanding of the genetic factors that underpin certain cancer risks, and our ability to perform a genetic test for the critical mutations. Angelina Jolie's attempt to surgically outwit her genetic legacy has not stopped at her breasts – last year she announced that she will also have her ovaries removed to protect her from ovarian cancer as well.

Naturally surgical intervention of this type is only feasible in the case of certain cancers. Removing ones breasts or ovaries is dramatic, but it is possible because in a woman beyond childbearing age these tissues are not functionally essential, and in the case of the breasts there are acceptable synthetic alternatives. Obviously, however, this approach isn't going to help much in the case of brain or lung cancer. Nevertheless the knowledge of mutations in a gene that alter ones susceptibility to develop cancer in certain environmental conditions could be extremely useful in guiding our behavioural choices. For example, if I was told that my genetic make up made the risk of developing brain cancer much greater if I drank coffee, I would be fairly certain that my frequent visits to Starbucks would be impacted. The number of potential gene by environment interactions is clearly enormous. However, despite the enormity of the task the dream of largely preventing cancer by matching genotype to lifestyle is alluring. Moreover, it is clear that this type of knowledge might not only be of use to the medical community in terms of preventative medicine. For example, if I was told that I had a genetic profile that was extremely resistant to developing lung cancer – even if I smoked, then I might not be concerned unduly about smoking the odd cigar or pack of cigarettes, visiting bars where I am exposed to passive smoking, or living in Beijing where I am exposed to high levels of air pollution. Personally I could benefit from the information, but I can imagine that this type of knowledge would also be of extreme interest to the tobacco industry, who might then be able to market their products towards sectors of the population where the negative effects of smoking are relatively small (or even absent).

This utopian personalised medical future however has apparently been dealt a severe blow in the last month by a paper published in the journal *Science*, by researchers Cristian Tomasetti and Bert Vogelstein from John Hopkins University in the USA. One aspect of cancer that is very evident but has defied explanation so far, is the fact that some types of cancer are very much more common than others. In part this can be rationalised by differences in the levels of exposure of different tissues to environmental factors. Hence cancers of the lung (7.6% lifetime risk of development) and bowel (7.2%) that are exposed directly to the environment, are among the more common types of cancer. But this explanation cannot explain why cancers of the prostate (13.2%), breast (12.9%) are more than 10x more likely than cancer of the liver (0.86%) and brain (0.81%) which are themselves massively more frequent than cancers of the pelvic bone (0.003%) and laryngeal cartilage (0.00072%). Similarly, different sections of the alimentary tract vary by more than a factor of 20 in their relative risk of developing cancer. For example cancer of the small intestine has a lifetime prevalence of 0.21% compared with the 7.2% risk for the large intestine. Tomasetti and Vogelstein had a simple hypothesis to explain this variation. Cancers are potentially formed when cells divide and mutations are introduced into the daughter cells. In theory then the rates of cancer development should depend on how many dividing cells there are in a tissue and how often they divide. Collecting data on the number of dividing stem cells in 31 different tissues they found a close correlation between this measure and the lifetime risk of developing cancer in that tissue. This correlation suggested that about 65% of the variation in cancer prevalence between tissues could be explained by stem cell division rates. In an online interview that followed the release of the paper Tomasetti has put this finding nicely into the context of the car accident analogy introduced earlier. Although the chance of having an accident depends on the road conditions (environmental factors) and the condition of the car (genetic factors), another major issue is the distance the car is driven. If you have a defective car and the environmental conditions are poor, your chance of an accident will still be extremely small if you only drive the car for a kilometre, compared to if you drive for 10000 kms. The distance travelled is equivalent to the rate of stem cell divisions.

Although some of the most prevalent types of cancer, like breast and prostate cancer, were not included into the analysis, the suggestion following publication of the work was that the model of cancer being caused by gene by environment interactions is no longer sufficient. Tomasetti is quoted as saying "If you go to the American Cancer Society web site and you check what are the causes of cancer, you will find a list of either inherited or environmental things. We are saying two-thirds (of the risk) is neither of them.". If this was true the dream that we might match together genotypes and environments to explain the majority of cancer risk is obviously not going to happen. Cancer by this model is mostly a lottery that you never want to win. This has some disturbing implications. In particular, if the majority of cancers are caused by chance, does it mean there is no point trying to prevent cancer by making appropriate lifestyle choices?

The paper unleased a surge of protest from groups interested in cancer prevention and subsequently there has been a subtle change in the language used to explain the results to the general public. Initially the 65% value was interpreted as meaning 65% of all cancers are caused by chance (see the quote above). Later on, however, this was modified to say that 65% of *the variation* in cancer rates *between tissues* is caused by the rate of stem cell divisions – which is what the paper actually showed. These sound similar, but are not the same thing. Hence if we consider just a single tissue like the lungs, it is true that your risk of developing lung cancer relative to pelvic bone cancer is much higher because lung tissue has a 2,000 times higher rate of stem cell divisions. But this rate of stem cell divisions in the lung is the same between me, you and everyone else. Using the car analogy again - as far as lung cancer is concerned we are all setting out on a 1000 km journey. So the question remains, would you rather go on that road trip choosing to drive only in the worst possible driving conditions and using a defective car, or would you prefer to do it in a well maintained car on days when the driving conditions are great. The answer is obvious, and that is why you have a 25 fold greater chance of developing lung cancer if you smoke than if you don't, and stopping smoking remains the best course of action to prevent lung cancer, irrespective of its risk relative to other cancer forms. In contrast, if it was shown that eating chocolate doubled my chance of pelvic bone cancer, I can be a bit more relaxed about that because pelvic bone cancer is just a 500m journey. So while cancer may have a component to it that is like a lottery that you don't ever want to win, it also remains a lottery that you can avoid winning by buying fewer tickets (making good lifestyle choices). The dream lives on.

When in Rome

March 2015

Here is a problem. Two people are walking down a road. One is a man and the other is a woman. The woman owns a small bag. Who is carrying the bag? It may come as a surprise to find that the solution to this problem depends on where you are located. In the west, of course, no man would be seen dead walking along carrying a woman's bag for her – unless it was a large heavy bag that she was struggling with. A small bag – never. You may be surprised to learn that in China it is really commonplace for a man to carry a woman's bag for her, even if it is really small. If the 'Beijinger' magazine is to be believed the number one complaint of young Chinese women who are dating western men is that they refuse to carry their bags for them!

There are lots of things that we do in modern society where there are a range of different possible solutions that would work equally well. Who carries a woman's small bag is but one example. Let's take another simple problem. There are six people in an elevator. They all intend to get out at the same floor. If they all attempted the leave at the same time there would be a jam in the elevator door. Who should leave the elevator first when the doors open? The solutions that are agreed by any particular group of people are arbitrary. We learn them when we are young. They are the conventions of our different societies. Normally we don't think about them. They are ingrained in us from an early age. They work because almost everyone in the society agrees that a given arbitrary solution is the 'correct' way to behave. In the case of the elevator problem, the solution in China is that the person who goes first is the one with the greatest status. Again, you may be surprised to find that this isn't a universal solution to the problem. In some societies it is the women who go first. The sum of these arbitrary behaviours is our culture. They ensure the smooth operation of our society. In some societies these rules take on incredible complexity and importance. Diplomatic incidents can easily be caused by culturally inappropriate behaviours.

We only really notice these things when we move to a different society where the cultural norms are different from what we are used to. When I first moved to China, I was constantly making small cultural mistakes because I was automatically behaving in the same way that I do in the west. The elevator exit problem was a regular case in point. I would often just expect my female PA to leave the elevator before me, so I would stand waiting when the doors opened. However, as her boss, she was waiting for me to leave first. More than once the doors started to close again before we resolved it. Overcoming our training and behaving in what we perceive is a culturally inappropriate way is difficult. It has taken me a while to become comfortable with marching out of the elevator first, leaving my PA to follow behind. This is not only a modern problem. There is a saying 'When in Rome', which is an abbreviation of 'When in Rome, do as the Romans do'. This saying is attributed to St Ambrose (who was born in 374 AD). Ambrose was the Bishop of Milan, in Italy, which at the time was a major city within the Roman Empire. Although he was based in Milan, Ambrose also had to attend the court of the Roman Emperor in Rome. This gave him a problem of deciding on which days to fast, and which days to pray. By convention these differed between the cities of Milan and Rome, but actually what was done when, was completely arbitrary. They were just different cultures. Ambrose recognised that the solution that caused the least conflict was to fit in with whatever was going on wherever you found yourself – hence the advice that remains today over 1600 years later 'When in Rome, do as the Romans do'. The cultural differences between China and the west are so great that whole books have been written on this issue, and after four years I am still very much learning what it means as far as 'When in Beijing' is concerned.

Some people moving to China from the west have been less able to cope with these cultural differences. Last year there was a letter to the international journal *Science* from a German researcher who had taken up a 1000 talents position somewhat like mine in Beijing. The main difference was that his position involved spending just 3 months per year in China. The German had resigned from his position because of a whole series of misunderstandings about the nature of his contract. Fundamentally, however, these misunderstandings seemed to me to mostly depend on a failure to adequately appreciate the differences in contracts between China and the west. These were combined with an unrealistic set of expectations about what life in China would be like. One complaint in the letter, for example, was that meetings in China had been conducted in Chinese, which the German didn't understand!

Culture, one might imagine is something uniquely human. However, an amazing experimental study recently published in the journal *Nature* has shown that in fact culture can be experimentally seeded into a population of small birds! To explain this remarkable experiment it is necessary to give you some information on the historical background that led up to the landmark study. This background starts in 1947 just after the second world war ended. At that time at an ornithologist called David Lack, who had just been appointed as the director of the Edward Grey Institute (EGI) of ornithology at Oxford University started a study of birds breeding in a wood just outside Oxford called Wytham wood.   To facilitate finding nests he installed nest boxes that were occupied primarily by small songbirds known as tits – principally great and blue tits. The study of the birds nesting in Wytham wood expanded and continued year after year until now the birds are some of the most well known in the world. Some of the birds have pedigrees of known parents that go back more than 35 generations! The primary interest of scientists working on these birds has been their population biology, and the effort over the sixty years since Lack initiated the study has generated more than 200 scientific papers. In the last decade this work has expanded under the guidance of the current director of the EGI (Professor Ben Sheldon) to study the social behaviour of the animals – mostly Great tits.

To evaluate the social behaviour the individual birds were fitted with small passive transponders known as PIT tags. The tags were attached to their legs. To tell which birds were associating with each other the researchers set up a set of feeding stations around the woods. When birds came to the feeders the tags on their legs were read and the time was recorded. So it was possible to work out which birds were travelling around feeding with each other.

Because there are a lot of marked birds in the woodland, and a lot of feeding stations the result has been the amalgamation of literally millions of recorded feeding events. Consequently, these birds have the one of the best documented social systems of any group of individuals on the planet. Among the more interesting aspects of this work has been the correlation between this social structure among the birds and their personalities. By bringing individual birds into the lab it was possible to look at how they explored a novel environment. Some birds were very bold and adventurous but others were more shy and reserved. What was interesting was when these personality types were overlaid onto the social structure of the birds society. It turned out that the boldest individuals had more social contacts and hence were at the central hubs of the societies. But the relationships formed by these birds were relatively shallow and more transient: particularly if they were interacting with other bold individuals. In contrast the shyer individuals tended to be on the periphery of the society, with fewer social contacts, but the contacts they had were more intense and longer lasting. Sound familiar? It seems that the social network of the great tits is not actually that far removed from our own. So if the birds have a society like ours do they have a culture as well. Sheldon decided to try and find out by doing a simple but extremely elegant experiment.

As I mentioned above culture is an arbitrary solution to a problem. To create an experimental culture Sheldon and colleagues brought individual great tits from different parts of the wood into captivity. These birds are extremely inquisitive and readily explore their environment. In captivity they were placed into a cage that had a small sliding door in one wall. The door was painted half red and half blue. For some birds from one part of the wood if they pushed the door to the left they got a small food reward from behind the door. For others, from a different part of the wood, they got a reward only if they pushed the door to the right. Great tits learn incredibly quickly how the door works, and which way to push it to get the reward. Once the birds had been trained to use the equipment with the sliding door in it they were released back into the wild. The researchers then put several pieces of the same equipment out into the field. But there was a twist, and this was the really clever part. With the equipment in the field it actually didn't matter which way the door was pushed. There was always a food reward behind it. But there were now two sets of birds that lived in different areas of the wood. One group had trained birds among them who had been trained to push the doors to the left, and the other group had individuals that had been trained to push the doors the other way. The question was, would two different cultures develop in the different parts of the wood – left pushers and right pushers, because individuals would learn from the trained individuals.

The answer was an unequivocal 'yes'. In the area where the left trained birds were released a left pushing culture developed. In contrast, where the right trained birds were released a right pushing culture emerged. In fact the two cultures were almost 95% consistent in how they pushed the door. Remember, it didn't actually matter which way the doors were pushed. The choice which way to push the door was completely arbitrary, and it depended completely on the individuals learning the arbitrary behaviour from the trained birds. As I mentioned above great tits are extremely inquisitive and behaviourally adventurous. Sooner or later in this system it was inevitable that a bird would discover that actually it doesn't matter which way the door was pushed. Actually this happened fairly quickly in both cultures. So why did the selective pushing of the door persist in the two cultures. If individuals worked out that the pushing direction didn't actually matter one might expect the consistency of behaviour would gradually get diluted and eliminated.   During the study the birds still carried the PIT tags from the previous study of their social structures, and there were readers at each piece of equipment it was possible to follow what happened with these individuals who had worked out how the equipment actually works.   What is amazing is that despite knowing that it didn't actually make any difference which way they pushed the door, all these individuals went through a phase of pushing both ways, but eventually they conformed to their own culture!   Social conformity it seems is important not only in our society, but in great tit societies as well. If that wasn't amazing enough occasionally birds would move from living with the left pushers to the area of the wood where the right pushers lived — and visa versa. What do you think happened in this situation?   The birds changed their behaviour! When in Rome the birds behaved like Romans!! Astonishing.

Irresistible drugs

April 2015

Plague is a rare disease. Worldwide there are about 2000 cases each year, almost all of which occur in Africa. In the UK nobody has died of plague for over 30 years. The same is true for most other European countries. Your chance of contracting plague in Europe is much lower than the risk of being struck by lightning. It hasn't always been like this. In 1347 for example there was an outbreak of plague that originated in Turkey and spread into Europe and North Africa. By 1351, four years after the initial outbreak,–it had killed a total of between 75 and 200 million people. This was somewhere between 30 to 60% of the total population at the time. Across Europe about one in every 2 people died! In southern Europe the death rate may have been as high as 80%. This was not an isolated outbreak. In fact plague recurred periodically across Europe every 10-30 years for hundreds of years, although not in quite so devastating a level as that observed in 1347-1351. During this period it was virtually impossible to live in Europe without your life being touched in some way by one of your close relatives dying of plague. If you look around old graveyards in the UK you can still find gravestones engraved with a skull and crossed bones to indicate that the occupants of the grave had died of plague, and they should not be disinterred.

Plague is transmitted by fleas that are carried by rats. Of course there are always rats around, and so too always fleas, so why did the plague erupt in such massive outbreaks interrupted by years when it was relatively unrecorded. According to a recent study it seems that in fact China was to blame! In a paper published in Proceedings of the National Academy of the USA last month Nils Stenseth and colleagues from the University of Oslo in Norway sought to find out what might be behind the plague outbreaks. They reasoned that it might have something to do with the climate. Warm wet spring and hot summers they argued might favour outbreaks of rats and enhance survival of their fleas which would then be more likely to pass on the plague. However this hypothesis proved to be incorrect – there was no link at all to the weather in Europe. Surprisingly however there was a link to the weather - in China! Several years prior to each major outbreak of plague in Europe there would be a period of mild weather in China. The authors speculated that the main reason for this linkage was that such periods favoured expansions in the populations of Chinese rodents in particular gerbils that also carry the plague and that such populations spread to the west when they are enlarged by successful periods of breeding.

Nowadays there are still warm periods in China and still the potential for expanding rodent populations to spread plague into Europe. There has been no plague outbreak however for more than 70 years. The reason is because of something remarkable that happened on the 28th of September 1928. That day a Scottish medic who worked at the University of London in England returned to his laboratory after being away on holiday. He was a bacteriologist who had been a medical officer during the first world war. On the battlefield he had noticed that many soldiers died, not because of the wounds they sustained in battle, but because their wounds became infected. Secondary infections actually killed more British soldiers than were directly killed by the enemy. After the war, and his discharge from the army, the medic decided to devote his life to the study of bacteriology and to try and develop ways of treating bacterial infection. He was, however, a rather untidy person and before leaving on holiday had left a whole stack of petri dishes with Staphylococcus growing on them just sitting on the lab bench. His intention was to clean them up when he came back from vacation. That morning he started to tidy up the plates when he noticed that on some of them a fungus had grown and around the fungus the bacteria had died. In fact this was not the first time that someone had observed this sort of negative interaction between fungi and bacteria. What sets this event apart however was the fact that the medic in question followed up the observation. He found out what the species of fungus was, and spent 12 years researching what the compound was that it produced that had such devastating effects on the bacteria. You all know this story already. The fungus was in the genus *Penicillium* and the medic was Alexander Fleming. His discovery of penicillin was hailed in 2000 as one of the top ten achievements made over the previous millenium, ranking alongside the invention of the computer and the internet.

For his discovery Fleming shared the 1945 Nobel prize for Medicine and Physiology with two other scientists, from Oxford, who were instrumental in isolating pure penicillin in 1940 and transforming it into a marketable drug (Chain and Flory). By 1950 Penicillin was in widespread mass production, and it completely changed our world. Bacterial infections which had dominated mortality in the period up to that point could now be effectively treated, and major killers like the plague, diphtheria, scarlet fever and sepsis became effectively things of the past. Several more antibiotics with similar or greater efficiency than penicillin, for example streptomycin, were discovered over the following decade, but then the discoveries trickled out, primarily because it was difficult to culture in the lab bacteria that could provide new sources of drugs, plus synthetic antibiotics were not easy to manufacture. The main issue was to develop drugs that could penetrate bacterial cell walls. This lack of new drugs wasn't an issue because at the time it really seemed that the war on infectious disease had been won. But there was a problem. Fleming noticed very early on that when colonies of bacteria were treated with penicillin their susceptibility to it varied. Some died very quickly, but others took much longer to succumb. They were 'resistant' to the drug. This phenomenon is why when taking a course of antibiotic drugs it is absolutely necessary to take the entire course to ensure every last bacterium is killed. If we don't do this then we effectively perform a selection experiment saving those bacteria that are most able to resist the drug. Fortunately we have sufficient antibiotics that if a pathogenic (disease causing) bacteria becomes resistant to one drug we can generally hit it with another one.

We normally think about evolution as taking many millions of years to happen, and in vertebrates like ourselves where a generation takes 20-30 years that is true. But in bacteria which have generation times measured in minutes evolution happens fast. In the last decade several strains of bacteria have emerged that show multiple antibiotic drug resistance, for example MRSA (Multiple drug resistant Staphylococcus aureus) and VISA (Vancomycin intermediate Staphylococcus aureus). The emergence of these and other antibiotic resistant strains combined with the fact we have developed only a trickle of new antibiotics (like Daptomycin and Linezolid) over the past 50 years, to which resistant bacteria have already developed, is one of, if not the single greatest threat to mankind. Forget terrorism. Forget climate change. They are completely trivial by comparison. If we don't solve this problem we will potentially be going back to the era when infectious disease killed most people before the age of 30! And this is not in some distant hypothetical future. Multiple drug resistant strains are already here and already causing severe problems in hospitals.

Two developments which have emerged in the last 6 months therefore are truly standout achievements that may allow us to postpone or even completely avert the impending disaster – the development of irresistible drugs. Not drugs that are irresistible to take, but hard to impossible for bacteria to develop resistance against. At a time when even rather minor scientific advances are hailed as major breakthroughs, these two things really do excel. The first is a paper in the journal Nature published in January this year (2015) by Kim Lewis and colleagues from a collaboration between several universities based in Boston, Massachusetts in the USA, and from Germany in Europe. This paper describes the discovery of a new drug called teixiobactin. Teixiobactin acts by attacking lipids in the cell walls of bacteria. This is its major strength and weakness. It is a strength because it is believed that antibiotic resistance develops largely because the existing antibiotics mostly disrupt proteins. Bacteria evolve to resist such antibiotics because proteins are part of the basic DNA-RNA-Protein mechanism. Hence it is relatively easy for bacteria to evolve new proteins that evade attack. Lipids on the other hand are less easily modified and hence evolution of resistance is a more difficult thing to achieve. Vancomycin which also attacks cell surface lipids was in use for 30 years before some resistance developed to it. In tests with teixiobactin the authors observed no evidence of bacterial resistance developing at all. This lends great hope that if it is eventually developed into a drug it will provide protection for a substantial period of time. By attacking lipids, however, Teixiobactin is only able to kill gram positive bacteria that do not have a protective outer membrane. The bacterium that produces Teixiobactin is a gram negative strain which is how it manages not to kill itself. This means it can kill MRSA and VISA, and a host of other resistant bacteria, which are the most immediate problems, but gram negative bacteria are outside its reach. Impressive as this is the paper in *Nature* is perhaps more noteworthy not because of the discovery of teixiobactin but the methodology that enabled its discovery. It has been known for some time that a major hindrance for the development of new antibiotics is that most bacteria, that may produce useful agents able to kill the bacteria we don't want, are almost impossible to culture in the laboratory. In fact almost 99% of known bacteria can't be cultured using conventional methods. To overcome this problem Lewis and colleagues devised an ingenious approach that like most brilliant things is simple and obvious when you know what it is. Reasoning that most bacteria fail in culture because we fail to provide them with optimal conditions including their required nutrition and relevant growth factors, they invented a plate that could be inoculated with soil bacteria and then buried back into soil where natural nutrients and growth factors could percolate in, and foster growth of bacteria inside. It is spectacularly successful, resulting in greater than 50% culture rate for soil bacteria compared to 1% in a standard petri dish. It's so simple that it's amazing it took half a decade for someone to think of it. Although the press have reported that this device was 'invented' in the Nature paper, in fact the authors published the method and device 5 years ago.

My only disappointment with this technology is that the authors chose to call this brilliant new device the "iChip". My heart sank when I read it. It seems nowadays that everything with any claim to genius has to be preceded by an "i" to emulate the brilliant iMac, iPhone, iPad etc originated by Steve Jobs and Apple. In so doing they instantly dilute its novelty and brilliance. There isn't even any great explanation why iChip is even a vaguely appropriate name – it is a contraction of Isolation Chip. The only use of a preceding "i" that I have seen by someone other than Jobs/Apple that I think is truly brilliant was that shortly after Steve Jobs died I saw a guy walking around Beijing with a home made T-shirt on that had a grave stone drawn on it with the simple epitaph "Steve Jobs. iDead"! Despite its name the hope in the future is that this methodology will allow us to discover many more novel antibiotics. If you are looking for somewhere to invest, then keep an eye on the company NovoBiotic pharmaceuticals based in Boston where the first author on the paper is based.

Will Teixiobactin prove to be the irresistible drug the authors promise? The jury is out. First the studies have only been performed in mice so far, and due to toxicity issues generally less than 10% of drugs that are effective in mice make it to the market. Plus recent drugs that were also considered irresistible, like daptomycin which was also isolated from a soil bacterium, led to resistant bacterial strains in under 3 years, when introduced into clinical practice. For linezolid the resistant strains emerged within a year of launch. Less widely reported in the press, the second advance is perhaps even more important then, and this is the development of a completely new type of antibiotic which is based on trying to mimic what happens in our own bodies when we attack invading bacteria. These drugs called 'defensin mimetics' are also very unlikely to spark resistance because they attack cell wall lipids, and one of them Brilacidin produced by the company Cellceutix is already in phase three of development. Unlike Teixiobactin it can kill both gram positive and gram negative bacteria. Plus being in phase 3 it is much closer to being an actual drug. Maybe the cavalry may have arrived just in time. Let's hope so.

How dogs make us fall in love with them

May 2015

We used to own a cat. His name was Freddy. He was black with a small white patch under his chin. I say that we owned him because he was actually a kitten born to another cat that we owned and so he was ours from the very second he was born. Yet like all cats we didn't really possess him. Freddy had a life of his own. He would come and go into our apartment via a cat flap in the door whenever he pleased. He would spend long hours away from our house doing things that we had no idea of. He often caught wild birds and mice, and could have easily sustained himself if we were to dry up as a source of food. In the middle of the night he would come home, sneak into our bed and warm his paws up on my wife's back. One day, without any warning, he disappeared and never came back. We searched the local roads to see if he had been hit by a car, but found nothing. We reported him missing to the police who also found nothing. His disappearance was a complete mystery – but I would not have been surprised if we had found out that actually Freddy was living a double life. He had probably befriended some old lady somewhere who was feeding him like we did, and let him sleep on her sofa by the fire, and one day maybe Freddy thought, actually I'll just stay here tonight – and from that point onwards he just didn't came back to us. That is why we never actually really owned him. Freddy owned himself and occasionally he graced us with his presence.

We also used to own a dog. By coincidence he was also completely black, but had a small white bit under his chin. His name was Basil. Basil we really owned. He was completely and utterly dependent on us. We could not trust him to leave the house on his own because he would just not have a clue how to behave outside. Basil was a member of our family in a way that Freddy never really was. When I wrote my first book I included basil among the 'people' it was dedicated to. I spent many hours teaching him tricks – how to shake paws, lie down, sit and wait. He would cock his head to one side and look into my eyes as I taught him, like he was really trying to understand what I was trying to teach him. In later life he bore with great patience our young children, and when he eventually died, when they were respectively 3 and 5 years old, it was a momentously sad day for our family. Now in their twenties my children say that Basil's death is about the only thing they remember from their very early childhood. Probably because it was the first time they ever saw their mother in tears. Why is it that dogs become such integral parts of our lives in a way that cats never seem to? For Basil we were the central components of his entire existence, and he in turn was a central component of our lives. For Freddy, we were two additional but somewhat incidental elements in his world, probably of equal significance to the sofa or his food bowl.

Now work from Japan, published in Science last month (April) is shedding light on this interspecies love affair. Lead researcher Takefumi Kikusui and colleagues from Azabu university studied how dogs and humans gaze at each other, and what the metabolic consequences are of such behaviour. Gazing is in fact an essential part of human to human social behaviour. Gazing into another persons eyes is one of the strongest forms of communication of intentions and affiliation. Studies in the 1990s showed that during one on one interactions mutual gaze conveys several important messages. People who avert their eyes during such contact – called 'gaze avoidance' are generally rated as more deceptive and less sincere. There are also some clear sex differences in how gaze avoidance is interpreted. Gaze avoiding women were rated as being disagreeable, unattractive, unconscientious and even of lower intelligence. In men gaze avoidance was seen as a predictor of psychological problems. In both sexes gaze avoidance is a symptom of autism spectrum disorder. Between a mother and her newborn infant, 'mutual gaze' is one of the most fundamental manifestations of social attachment. Last year a paper in Brain Research from scientists at Baylor College of medicine in the USA, showed that a key component of this gazing behaviour of mothers towards their children was mediated via the hormone oxytocin. Oxytocin has been known for decades to be an important hormone that is involved in the process of giving birth, and also in ejecting milk from the nipple. However in the last decade it has become clear that oxytocin is also an important mediator of social behaviours, as well as social trust, emotional recognition and empathy. The particular experiment published in Brain Research involved filming mothers interacting with their babies over a 50 minute period, which included a phase of interaction when the mothers were instructed to maintain a neutral expression. This process is known to induce mild stress in infants. During the periods prior to and after the stress period the observers recorded how long mothers maintained gaze with their babies and also how frequently they broke off gazing. They showed that these two behaviours were closely linked to the increase in maternal oxytocin levels during the 50 minute interaction. When mothers had a high oxytocin response, they also showed longer gazing and less frequent break-offs in their gazing behaviour. This is important because quality provision of maternal care and the formation of strong maternal-infant attachment bonds have long term effects on child development. Sensitive and responsive maternal behaviour has direct effects on a child's life long capacity for social adaptation and stress regulation.

In fact oxytocin may also be a critical feature for adult-adult relationships as well as maternal-offspring bonding. Our understanding of this link has been greatly facilitated by studies of two vole species in the USA. (A vole is a small mammal about the same size as a mouse). One type of vole – the prairie vole is monogamous (adults mate almost exclusively with a single partner), forming enduring and stable pair bonds with a single partner, and typically the pair show co-operative parental behaviour in the care of their offspring. In nature the majority of prairie voles that lose their partner never take on another! In contrast, montane voles and meadow voles, are not monogamous and normally show no co-operative care of their offspring. These species differences can be studied in the laboratory using various paradigms which measure the fidelity of a vole when given opportunities to consort with a stranger. It turns out prairie voles seldom cheat in such experiments – while montane and meadow voles are not so loyal. Two neuropeptides (oxytocin and vasopressin) have been linked to these differences, based on different levels of these hormones and their receptors in the brain between the different species. This correlation doesn't prove anything, but giving prairie voles oxytocin or vasopressin increases their likelihood to form a strong bond, while giving them chemicals that block these hormones turns them into cheats.

Oxytocin it seems is more important for forming pair bonds in females and vasopressin more important in males. So experimentally the evidence of their involvement is quite strong. You are probably wondering now whether these same hormones mediate the pair bond in humans? We know that oxytocin levels are elevated during orgasm in women and vasopressin levels increase in men during sexual arousal. Also when humans are shown pictures of people they are in love with, functional magnetic resonance imaging of their brains show elevated blood flow in regions of the brain that are rich in oxytocin, vasopressin and their respective receptors. So the fundamental basics of the system are in place in humans as well. If you are thinking of getting married soon then maybe you should think about getting your partners oxytocin and/or vasopressin levels measured after they look at pictures of you!

It seems then that oxytocin forms a vital element in the formation of bonds between mothers and their children and between sexual partners. The recent paper in *Science* concerning dogs and their owners gazing at each other suggests that dogs have been able to hijack this pathway to make us fall in love with them as well. It is actually already known that when humans and dogs interact with each other by patting, stroking and sitting in contact with one another there is a mutual increase in oxytocin levels. Moreover, when a dog gazes into its owners eyes the same Japanese researchers had previously shown that there is an increase in the owners oxytocin levels, which did not occur if the gaze was blocked. What the latest paper shows is that in fact humans and dogs are engaged in a positive loop of reinforcement when they gaze into each others eyes.

First, they examined whether the level of dog-human mutual gazing behaviour affected the oxytocin concentrations of both participants in the interaction. Oxytocin levels were measured in urine before and after a 30 minute interaction during which the researchers recorded the amount of time that the owners gazed with their dogs, spoke to them or touched them. The most significant factor influencing the change in oxytocin levels of the owners was the duration of the mutual gazing. Similarly the highest change in the levels of oxytocin were in the dogs where the mutual gazing was the longest. When a dog looks into your eyes it is stimulating production of a powerful hormone in you that strengthens the bond between you and it.   But equally your gaze stimulates the same hormone in your dog. It is a mutual stimulatory event – in a very real sense you are both falling in love. Interestingly the researchers performed the same experiment with tame wolves and their owners and found no such relationship. Consequently the researchers suggested that the mutual gazing between a human and a dog must have evolved during the process of domestication. Indeed the ability of a dog to tap into the basic pathway of human emotions may have been an instrumental aspect of the domestication process in the first place.

One issue is that these observations are, like the associations between oxytocin and pair bonds in different vole species, just correlations. Can we do some experiments like the vole biologists did to prove that oxytocin levels are causally implicated in these links? This is what the researchers from Japan did in a second experiment in their Science paper. Oxytocin can be administered to dogs via a nasal spray. So the researchers did this to dogs and then allowed them to interact with their owners and two strangers. Giving the dogs oxytocin (compared to just saline solution) increased the time they spent gazing at their owners and not the strangers. Plus the urine concentration of oxytocin in the owners increased more as well when their dogs were experimentally treated with oxytocin. But the effect only occurred in female dogs. In male dogs there was no effect on the dog gazing, and no impact on the owner oxytocin levels either. However, the researchers did not study the potential role of vasopressin, which as noted above is more important in male vole pair bonding than oxytocin. Perhaps the difference between male and female dogs in this experiment was just that male dogs have hijacked the vasopressin arm of the system and female dogs the oxytocin side.

It seems that this is really something unique about the dog-human relationship. Our closest genetic relative, the Chimpanzee, does not elicit the same response. Moreover, the closest relative to the dog, the wolf, also is unable to tap into this system as well. Humans and dogs are locked into a positive feedback loop mediated via their mutual gazing behaviour that has many similarities to the human-human pair bonds that underpin long term sexual liaisons and mother-infant interactions. During domestication dogs hijacked a deep rooted and fundamental component of the physiology that underpins our social behaviour. That's why they become members of our families in a way that cats never do. When a cat looks into your eyes it is probably because your head is in the way of something else it is trying to look at. When a dog looks into your eyes you are falling in love with it, and it with you.

Chocolate for weight loss

June 2015

We have a family friend in the UK, let's call her Claire, who has always struggled with body weight issues. Claire is almost perpetually on a diet. Whatever the latest diet fad is, Claire has tried it. She is not alone in her obsession with weight loss and dieting. Losing weight by dieting is a popular pastime, particularly in the United States. Estimates by the Boston Medical Center suggest that about 25% of Americans engage in a diet at some point each year, and 33 billion US dollars are spent on weight loss products in America annually. As obesity spreads around the world more and more people will be turning to dieting as a method to reverse the weight they have gained. Dieting to lose weight is fundamentally all about calories. You need to consume fewer calories than you expend so that the balance must be taken out of your fat stores. It is easiest to do this if you cut out foods that contain high levels of fat and sugar. But if that is only what you do, then it is also very easy to fail because these foods are also the most tasty and pleasurable to eat. A whole industry has grown up around the ideas of what the optimal combination of foods might be to enable you to lose weight, but also enable you to stay on the diet for a long time, and hence maximise your long term weight loss. High fiber and high protein diets (e.g. Atkins type diets) both seem to have some success in this respect. But there are always new ideas being published almost monthly.

In March this year there was a paper published in the

International Archives of Medicine, which came up with an interesting suggestion on how to succeed in your dieting, and the answer was a bit unexpected. Eat chocolate. In particular eat a bar of dark chocolate every day while you are dieting! The findings were splashed across the front page of 'Das Bild', a German newspaper, which is actually the largest circulation newspaper in Europe. Several other daily papers and numerous diet and nutrition magazines ran the story. Typical headlines were "Why you must eat chocolate daily" (Shape magazine), "Chocolate accelerates weight loss" (UK Daily express) and "Good news for slimmers! Eating chocolate can help you lose weight" (UK Daily Star). I didn't see Claire for the last couple of months but it wouldn't surprise me to find that she is eating a bar of chocolate every day now in response. After all this seems too good to be true – chocolate is one of the foods that is *always* bad for you if you are dieting. It is almost only sugar and fat, and a typical bar contains about 200-300 kcals which is about 10% of your total daily energy needs. How could it be that chocolate has now become a diet food?

The work was conducted at the Institute of Diet and Health in Germany. The study involved 3 groups of people who were allocated at random to 3 different treatments. One group was a control group. The second and third groups were put onto a standard low-carbohydrate diet but one of these 'diet' groups was also asked to eat a 45g bar of dark chocolate every day. The individuals were followed for 3 weeks. At the end of that time both of the diet groups had lost about 5 pounds in weight relative to the controls, but the individuals in the diet plus chocolate group had lost 10% more weight than the diet only group. The results were statistically significant, meaning the probability that the result was just a chance difference was less than 5%. It seems clear cut enough to justify the big headlines and I have no doubt Claire and probably hundreds of thousands of others took up the messages in the headlines with glee.

If she did I hope she stops soon, because the authors of the too good to be true chocolate diet paper have just (May 2015) revealed that the experiment was in fact a hoax. The first author on the paper and 'Director' of the Institute of Diet and Nutrition was actually a journalist called John Bohannon, and the 'Institute' is just a website. His collaborators on the paper included a doctor and a TV reporter. So why would someone do that? The motivation behind Bohannon's hoax experiment was to reveal just how easy it is to do a piece of rubbish scientific research, get it published in a scientific journal and then get the science press to promote the story to the general public.

The interesting thing is that the results of the experiment weren't just made up. They actually did an experiment. They advertised on the internet and got 16 volunteers to take part. They then took some blood samples and administered a set of questionnaires, and randomly allocated them to the three different groups. After 3 weeks they then brought them back, did some more blood testing and questionnaires and sent them on their way. And sure enough the results for weight loss were statistically significant at the 5% level after rejecting the results from one subject as an 'outlier'. So what is the problem? If it was a legitimate study is thatnot OK? The main issue is that they actually measured a whole heap of things – including weight loss, but also cholesterol levels and various other blood biochemicals and numerous things in the questionnaire. In fact there were 18 parameters measured in total. As Bohannan has pointed out, in the article where the hoax is revealed, the more things you measure the more likely it is you will get at least one thing significant. They knew with a small sample size and a bit of judicious data rejection (ie one outlier) by measuring so many things at least one thing would end up significant just by chance. They didn't mind what it was. The story could equally have been 'chocolate improves your blood glucose levels', or 'chocolate reduces your blood pressure'. As it turned out the weight loss difference was significant so they went with that as the key finding.

The next step was to get it published in a scientific journal. The global repository of scientific knowledge is protected by the process of peer review. Before a paper is published in a scientific journal it is scrutinised by several 'reviewers' to pick up any problems: like having a really small sample size and rejecting data to make things significant! How was it possible given these obvious flaws to get the paper published in a scientific journal? Why did the reviewers not pick up the issues? The answer is that in the last decade the world of scientific publishing has become invaded by hundreds of scientific journals that are basically just 'pay to publish' outlets. You pay the journal a fee and within a few weeks the paper is published without any actual peer review. Bohannon has a history of interaction with this world of dubious scientific publishing because last year he submitted a completely made up and blatantly flawed paper to over 5000 such journals and had it accepted at more than half of them. So he selected twenty of these journals and claims that within 24h he already had several acceptances to publish. He eventually chose the International Archives of Medicine. Here Bohannons story of what happened diverges from the journals account. Bohannon says that the editor emailed him to let him know that they had produced an "outstanding manuscript," and that for just 600 Euros it "could be accepted directly in our premier journal." And despite the journal website claiming that all papers are rigorously peer reviewed it was published within 2 weeks of the payment being taken, without changes to a single word. If you check on the journal web site you get a rather different story. There the editors claim they made a simple mistake. The editor mailed Bohannon, he says, to tell him the paper had been 'accepted to go for peer review', but an editorial assistant who was copied into that email misread the message and thought it had already been peer reviewed and accidentally posted it online. Once they discovered the mistake they took the article down (within 3 days) and then rejected it. At this point it is hard to know who is telling the truth.

The last step of the scam was to disseminate the story to the scientific press via a press release. They distributed this to 'Newswise' a site that aggregates press releases and distributes them to >5000 journalists. It is an interesting story and given the huge interest in dieting not surprisingly some of them took it up. Bohannon cites this as evidence that science reporters in general are lazy and just pick up press releases without any critical appraisal of their merit and regurgitate them unquestioningly to the public.

This whole episode raises a number of important issues about the way science is performed, published and how it is disseminated to the public. Can we really trust scientists to do good work without fishing for p-values in dubious sample sizes? Do journals really publish anything as long as payment is made? Can we trust science journalists to check the facts before releasing the news on an unsuspecting public? Bohannon's experiment seems to suggest the answer to all these questions is a resounding NO. But is it really?

There are several major problems with what Bohannon and his colleagues did. The first is that they ran an unregistered clinical trial. Doing research on other human beings has a long, and not so glorious, history, and so in the 1950s procedures were brought into place to make sure that anyone wanting to run an experiment on human subjects must get the procedures for the experiment ethically reviewed and have the trial registered. If Bohannon had done that, the trial itself would almost certainly not have been permitted to proceed, because it is simply not allowed to go on a fishing trip like the one he describes, to see if something comes up as significant. Studies need defined primary goals that are put forward in advance of the trial being started. Moreover, they have to recruit sufficient subjects to detect given effect sizes for these primary outcomes based on a statistical power analysis. A trial to find an effect of chocolate on weight loss with 16 subjects would almost certainly not make the grade. Finally ethical review boards are set up by established legitimate research institutions to monitor research activity on their own sites. The reason Bohannon couldn't go through this step is because such a board would instantly recognise that his bogus Institute was indeed just that. In the real world, not populated by investigative journalists who can run roughshod over the rules, such a trial would never have been given ethical permission to be conducted.

What actually happened at the scientific journal is disputed (see above). Nevertheless it is clear that there are lots of 'scientific' journals out there that do not adhere to peer review and publish whatever comes in, as long as it comes with the appropriate fees. Most non-scientists are unaware of this problem, but knowledge of it is becoming more widespread and Bohannon has been in the vanguard of exposing these dubious practices. Unfortunately there are now thousands of these journals and enough people with more money than ability to populate their pages with garbage science. In this specific instance it is hard to know who is telling the truth, but it exposes a much wider problem. Bohannon would never have got this paper through peer review in a decent journal, which is why he didn't send it there. The problem is how someone who is not a scientist is to know the difference between say "The International Archives of Medicine" and "The New England Journal of Medicine", one of which is an open access pay for publish outlet, and the other is the premier medical journal in the world. The only reason this system persists is because scientists themselves benefit from publishing their work in such non-peer reviewed outlets. If evaluation committees started to view negatively publications in these places then scientists would pretty quickly learn that it is in their best interests to avoid them.

The final issue is how science journalists pick up and do not critically review the information in press releases before selling it to the public. Scientists are unfortunately forced into this world by the necessity to show their studies have 'impact', combined with the almost limitless demand for 'news'. Actually in this case the numbers tell a rather interesting story. Bohannon claims to have exposed lazy journalism. The paper was sent to 'Newswise' which sent the release to over 5000 journalists. However, it was only picked up and run by 13 of them. So while Bohannon published the revelation of his hoax under the headline 'how I fooled millions' in fact the numbers suggest that 99.9% of the journalists who got the story didn't go with it. Potentially because they evaluated its content, and perhaps the actual paper before it was taken down by the journal, and decided it wasn't sound enough to base a story on. Maybe the system isn't so bad after all. The problem is that 13 big news stories and follow on internet coverage is probably enough to get hundreds of thousands of people like my friend Claire eating chocolate to help their diets.

So this comes to the final problem. Bohannon claims to be exposing dubious practices in science. But is it really ethically acceptable for someone to run an unregistered clinical trial, particularly one including blood sampling. Then take what are known to be spurious results (and if we take the journals version of events data rejected for publication) and promote it to the press, with an outcome that could potentially damage peoples health? Who is really in the wrong here? Is this a story of dubious science, or ethically inappropriate journalism?

Pandas are cool – its official!

July 2015

There are few animals on the planet as familiar to people as the Giant Panda. Its role as the logo for the world wide fund for nature, the perilous nature of its existence in the wild where less than 2000 are still alive, its seemingly pathological reluctance to reproduce, and the fact that captive animals have been exported worldwide as a symbol of Chinese political friendship for decades, continue to sustain its iconic status. Yet, despite this phenomenal popularity and familiarity, the political and conservation status of the panda makes doing research work on them particularly difficult. There are many, and justified, hurdles to jump to do scientific work on animals: to do research on pandas those hurdles are both more abundant and substantially higher. Therefore many fundamental measurements that have been made on other species are simply lacking for this enigmatic bear. We are starting, however, to fill in these gaps, some of which are detailed in a paper published this month in *Science* (July 2015), which summarises some work we have been doing on the energy metabolism of these amazing animals. This work was a collaboration between my own group at the Institute of Genetics and Developmental Biology, at the Chinese Academy of Sciences (CAS) in Beijing, the group led by Professor Wei Fuwen from the Institute of Zoology at CAS also in Beijing, and the Beijing zoo.

Perhaps one bit of giant panda biology almost everyone knows is that it is a carnivore that became a vegetarian. The mammalian order Carnivora includes several families of animals including the canids (wolves, dogs and foxes), felids (cats such as the lion and tiger), mustelids (weasels etc), pinnipeds (seals, walruses and sea-lions) and ursids (bears). All these groups, apart from the bears, subsist almost exclusively by killing and eating other animals. Because meat is easily digested they are characterised by having short and simple alimentary tracts. A domestic cat, for example, has an alimentary tract that is not much longer than that of a mouse. Apart from the polar bear, other bears include various amounts of vegetable material into their diets. Brown bears in the USA, for example, exploit seasonally available berries. The panda is a bear that has taken this habit to its ultimate extreme: eating almost only bamboo. It seems from the fossil record that they first started bamboo eating around 7 million years ago, but became completely herbivorous much more recently (about 2.2 million years ago). Interestingly, pandas have a mutation in the umami taste receptor, which is responsible in part for the taste of meat. It is disputed what came first the loss of the taste for meat or the vegetarian diet. Some have suggested that the mutation may actually have been the driving force that led them to specialise on a vegetarian diet. It seems likely, however, that eating bamboo preceded the loss of function mutation in umami receptors, because loss of the ability to taste meat would presumably be disadvantageous in a meat eater, and the mutation would have been eliminated. In contrast, loss of ability to taste meat in a vegetarian would probably be selectively neutral, and could then spread through the population by the neutral process of genetic drift. Such a drift process has probably played a large role in shaping the panda genome, which has apparently always had a very small effective population size (which favours genetic drift). Supporting this suggestion estimates of when the umami receptor mutation happened are around 4.5 million years ago, substantially after the fossil evidence that the animals started to eat bamboo. Although pandas have many adaptations for eating bamboo (like an extra 'thumb' to help hold the shoots), and the large jaw muscles that enable them to break into tough bamboo stems, that give the panda its characteristic rounded face, these adaptations do not include a modified alimentary tract. The panda has the guts of a lion: ideal for digesting meat, but very inefficient for digesting bamboo. So they have to eat lots of it, and it passes almost unscathed through their digestive system with very little of it being absorbed into their bodies. They need to eat perhaps as much as 10-20 kgs per day. Eating that much bamboo takes up a big part of the pandas day.

Panda at Beijing zoo eating bamboo

Scientists have long speculated that to survive on such a low quality food pandas must have a low rate of metabolism. However, until our paper in *Science* nobody had yet managed to measure exactly how much energy they use. We applied a technique called the doubly-labelled water method which measures the rate at which animals eliminate stable isotopes from their bodies to measure the metabolism of 5 captive pandas in the Beijing zoo, and 3 wild pandas living in the Foping nature reserve in Shaanxi province – near the city of Xi'an where the terracotta warriors live. The answer it turns out is that their metabolic rate is exceptionally low. Corrected for their body weight (which for the animals we measured averaged 92 kg) the panda has substantially lower metabolism than almost all the other mammals. This includes all the ones that we often think of as having low metabolic rates – like the koala and echidna. In relative terms it is about exactly the same level as the 3 toed sloth. In fact the metabolism of the panda is closer to what would be predicted for a 90 kg reptile, than the prediction for a 90 kg mammal. We often think of modern day humans as having low metabolic rates because of our 'sloth like' lifestyles, but compared to a panda (and the actual sloth) we are like supercharged racing cars. We found that across all 8 measurements the pandas expended just 5.2 megajoules of energy per day (1250 calories). A 90 kg human living in modern society by comparison expends about 12 megajoules per day (3000 calories)!

To cross check the estimated metabolic rates we took advantage of the fact that almost every day the animal handlers and researchers at Beijing zoo collect the feces that the pandas produce and weigh them. We therefore did an experiment to find the relationship between how much food a panda eats and how much feces it produces, and from that relationship we were able to work back from the weighed fecal records to find out exactly how much food the pandas had eaten every day, through a whole annual cycle, and how much of the energy from that food was supplying their metabolism. The two answers – from the food intake and the isotope method didn't match up exactly. We showed that this was probably because the food intake method doesn't take into account energy that is lost in urine. When we checked the energy content in the urine we were really surprised to find that over 20% of the energy that the panda eats appears in its urine. This is exceptionally high. In mice for example the urinary energy is only 3%. Why is that? We already know that pandas use their urine to scent mark the areas where they live and to do that they secrete a variety of fatty acids and other volatile compounds into their urine. Perhaps that is why their urine contains so much energy. Nevertheless, it seems strange that an animal we regard as energetically on the edge can afford to put so much of its precious energy intake into this use – which in turn highlights the probable importance of this secretion and such scent marking in their daily lives.

How pandas achieve such low rates of energy use was the focus of the second part of our paper. Much of the energy that our bodies use is burned up in relatively few 'high power' organs – like our brains, kidneys, heart and liver. Using historical autopsy data we found that pandas have particularly small organs for their body size. Their brains are only 82% of the expected size, their kidneys only 74.5% and their livers a remarkable 62.8% of the expected size for a 90kg mammal. Plus if you ever went to see a panda in a zoo you will know that they are not the most active of animals. In the wild they are more active – but not much. Indeed using GPS loggers we found that in the wild pandas move on average at a leisurely average speed of just 26.9 metres per HOUR! On a typical day in the wild a panda may move less then 200m.

A key physiological system that is involved in regulating our metabolism is the thyroid hormone system. We suspected given their very low metabolic rates that pandas might have something unusual going on with their thyroid hormones and this hunch turned out to be correct. Pandas have very low levels of the main thyroid hormones T4 and T3. In fact an awake panda has thyroid hormone levels that are even lower than in a hibernating black bear, which has its metabolism almost completely switched off. We explored the panda genome to see if there were distinctive features of the thyroid signalling and production pathways in pandas when compared with humans, mice and several other Carnivorans. We found that pandas do have mutations in the genes DIO2 and DIO1 that interconvert the different thyroid hormones that were not present in humans or mice. But these mutations were also present in their Carnivoran relatives so probably don't play a key role in their low thyroid hormone levels. In addition, however we did find a panda unique mutation in a critical gene (Dual oxidase 2: DUOX2) that is involved in thyroid hormone synthesis. The DUOX2 gene catalyses the production of hydrogen peroxide that is used in the last step of thyroid hormone production. It is a very large gene comprising in humans 32 exons. We found in the panda that in exon 16 there is a single nucleotide mutation that creates a premature stop codon that probably results in a truncated and potentially non-functional DUOX2 protein, reducing the efficiency of thyroid hormone production. Supporting this idea, natural mutants of DUOX2 in humans, and mice with the gene deliberately knocked out, both show extremely low thyroid hormone levels.

People who have low levels of thyroid levels often complain that they feel cold. This is potentially because their lowered metabolic rate is insufficient to keep them warm. Having such a low metabolism raises a similar problem for the panda – how does it manage to keep warm. The answer is that despite living in a semi-tropical habitat the panda has a really thick fur coat. This serves to trap what little heat their metabolism produces inside their bodies to maintain their body temperature. A direct consequence of which is that the surface temperature of the panda (measured using a thermal imaging camera) is about 10 ^{o}C cooler than the surface of other black and white animals like the zebra, Dalmatian dog or the Holstein dairy cow, when measured at the same air temperature. Pandas it seems are literally cool! Incidentally I also had a pretty cool experience while we were doing this work. One day at the zoo I was feeding one of the pandas some bamboo and I got distracted by something and the panda reached through the bars and took a swipe at me to get the bamboo I was holding. But he missed and took a small chunk out of the rather expensive leather jacket I was wearing. Normally I would have been very annoyed by this but then I thought, if someone asks how I got the damage to my coat I could in all honesty say "Actually I got it in a close encounter with a giant panda". Surely, there can be no cooler answer!

Mathematically the most attractive.

August 2015

Plato, the ancient Greek philosopher said that "beauty lies in the eye of the beholder", emphasising that our views of who is beautiful are all individually variable. However, look at the following two images. Which do you find more attractive?

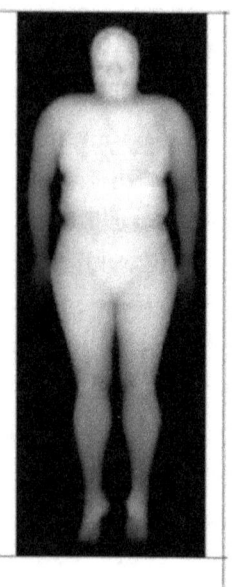 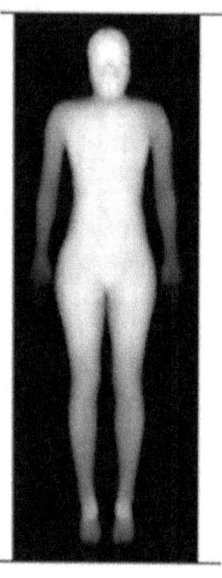

Chances are, independent of whether you are male or female, Asian, Caucasian or African, you chose the image on the right. In fact less than 1 in a hundred people think the image on the left is more attractive. Beauty may lie in the eye of the beholder, but beholders almost unanimously agree that the image on the right is more attractive. Why is that the case? There are several very clear differences between the two images. In particular, the image on the right has a very distinct waist that is considerably narrower than her hips. If you calculate the ratio of the waist to the hips (called the waist to hip ratio or WHR), then for the individual on the left the WHR is 0.82 and for the image on the right it is 0.68. In the 1990s it was widely suggested that waist to hip ratio is a key factor that drives our impressions of physical attractiveness in females, with a value of around 0.7 being most favoured. It was noted at the time that the waist to hip ratios of females that appear in pornography magazines almost all have WHRs in the range 0.65 to 0.7 and that this had been constant for almost 50 years. This role of WHR in perceptions of beauty has also been demonstrated experimentally by surgically manipulating WHR and showing that when WHR is surgically changed to around 0.7, individuals are rated as more attractive. Some studies suggested that this might be because females with WHR values around 0.7 are more healthy, and one study even suggested that women with WHRs of 0.7 are smarter, than individuals with higher WHRs. Yet other studies suggested that females with lower WHRs have more sons, although this was later shown to be incorrect. These arguments implicitly suggested therefore that the reason we find certain body shapes attractive and others less attractive is that we are tuned in to what the consequences of these body shapes are for evolutionary fitness. By this argument WHR is an honest marker for evolutionary fitness. We might think we are looking at a thin waist but what our evolved brains see is how healthy the person is and how many children she could have.

In the late 1990s, however, it was recognised that the main reason why the waist to hip ratio differs between individuals is primarily because of changes in the size of the waist, not the hips: and this is linked to differences in obesity. As people become more obese they deposit fat predominantly in the abdomen compared to the hips, and this results in the waist to hip ratio getting larger. So the individual on the left in the images above not only has a larger waist to hip ratio but is also more obese. The individual on the left has 38% body fat and the one on the right has only 23% body fat. In terms of body mass index (weight divided by height^2) which is a widely used measure of obesity, the left image has a BMI of 30 and would be classed as 'obese' while the one on the right has a BMI of 19, and would be classed at the lower end of the 'normal' range. So what is more important for our impressions of attractiveness – the waist to hip ratio or the level of obesity. To find out one needs to do a set of studies where the WHR and obesity vary independently of each other. Image sets were generated that broke the link of obesity to WHR, or even made the link artificially negative, and a whole series of studies in the late 1990s and early 2000s led primarily by Dr Martin Tovee at the University of Newcastle in the UK showed conclusively that it was the level of obesity that was being used to rate physical attractiveness, and the association with WHR only came about because generally individuals with low levels of obesity are those with low waist to hip ratios.

The American socialite Wallis Simpson famously said about female attractiveness in the 1930s that 'you can never be too thin or too rich'. But she was at least 50% wrong because one only needs to see the victims of famine to realise that you really can be too thin. Somewhere between morbid obesity and famine emaciation there is an optimum level of fatness that maximises physical attractiveness. A key question is what is that optimum level and why? Does the optimum correspond to the level that maximises evolutionary fitness?

For the last few years at the Chinese Academy of Sciences, Institute of Genetics and Developmental Biology in Beijing, my group has led an international collaboration involving research partners in 10 different countries trying to find out what the optimum fatness level is that makes females maximally attractive, and how this varies between different cultures. In particular we have been interested to find out whether the maximal level of physical attractiveness corresponds to the level of fatness that maximises evolutionary fitness. Fitness in evolutionary terms comprises two things: survival and fecundity. We were interested to know if when we look at someone and think they are physically attractive, are we actually making that assessment based on an evolutionary understanding of their potential for future survival and fecundity. The work was published this month in the journal PeerJ.

To test the idea that there is an optimum female body fatness, because it signals evolutionary fitness, we built a predictive mathematical model which combined together the relationships between levels of obesity and the future risk of mortality from all causes, and the relationship between obesity and the future possibility of having children. This 'fitness' model predicts that the most physically attractive females should have a body mass index between 22.5 and 23.2, and that this optimum should be largely independent of ethnic background (we built separate models for

Caucasians, Africans and Asians – which comprise about 91% of humans currently living on earth, and they all gave similar answers). There is, however, another factor that might be important, and that is in our evolutionary history it has been suggested that we had to endure periodic famines. People who store more fat could potentially survive famines longer than individuals with lower fat reserves. So it might be attractive to be a little fatter than the value based only on current rates of survival and fecundity. Building this factor into the mathematical model suggested that the optimum fatness for maximum attractiveness should be at a BMI between 24 and 24.8. We might anticipate that populations that had experienced famine very recently would maybe be more sensitive to this factor than those populations where famine had not been experienced for hundreds of years. So for those populations that experienced recent famine the mean might be around 24.5 while those that had not experienced famine recently the mean might be around 23.

To test this idea we showed images of 21 females of different levels of fatness (controlled for their waist to hip ratio) to over 1300 people. The two images above are from this series of 21 images. The images came from a previous study conducted in the USA by Dr Mark Faries from the Stephen Austin State University in Texas. The raters were both males and females, and they came from three Asian populations (China, Iran and Mauritius), three Caucasian countries (the UK, Austria and Lithuania) and four African countries (Kenya, Nigeria, Morocco and Senegal). Perhaps unsurprisingly, in all the populations, males and females rated physical attractiveness of the female images very similarly. In other words what males in a given population regard as attractive or unattractive maps almost perfectly onto what females regard as attractive and unattractive in the same population. However, the highest levels of attractiveness did not correspond at all to our predictions from the evolutionary model. Indeed across the range of images we used (with BMIs from 19 to 34) there was a linear negative relationship between attractiveness and fatness. The thinner the image the more attractive it was rated, and that was true in all of the populations we studied. The most attractive image had a BMI of 19 (it was the image on the right above). Surprisingly, given the previous work which had suggested WHR was not important, we found some significant differences in the role of WHR between the different cultures. In particular in the Asian and Caucasian populations WHR was not important and ratings were based almost completely on the level of body fatness (or BMI). However, in the African populations there was also a larger and more significant independent effect of WHR. All populations we studied favoured thinner subjects (down to a BMI of 19) but in the African populations a more curvaceous body shape at any particular level of fatness was also regarded as more attractive.

The difference between the predicted peak in the relationship between attractiveness and BMI from our evolutionary mathematical model (BMI between 22.3 and 24.8 depending on the population and assumptions) and the actual levels of maximal attractiveness (BMI around 19) is enormous. For a woman of average height (1.55m) it is a difference in body weight of about 10 kg. Why did the model perform so poorly? One explanation might be that all societies are now exposed to a sort of 'Hollywood' ideal of attractiveness which emphasises thinness, and they rated the images we showed them accordingly. There is however a chicken and egg problem here. It is reasonable to ask: why the Hollywood ideal is so thin? Does the ideal follow on from the fact people find thinness attractive, or do people's ratings of attractiveness stem from the ideal? This doesn't actually explain anything, but simply changes the question to: why does the 'Hollywood' ideal not conform to the evolutionary model we built? Plus we did detect differences between the populations in the role of WHR suggesting that these populations had not all been homogenised by exposure to a 'Hollywood standard'.

There is a much better explanation for the discrepancy, and that is because we did not tell the raters how old the people in the images were, but rather we assumed the images were all the same age, and thought that the raters would do the same. However, when we asked the raters how old the people in the images were, there was a strong relationship between the estimated age of the person in the image and their BMI (but no relationship between the estimated age and their actual age). Have another look at the images above. How old do you think they are? In our sample people estimated the age of the subject on the left was 45, while the subject on the right was estimated to be 20. In fact their actual ages were 28 and 22. This association between BMI and estimated age is important because age is a very strong predictor of fecundity and survival. There is a peaked relationship between age and fecundity, with the maximum around the age of 18-22. In part this is because of the decline in ovarian reserve as females get older. However, the peak occurs because females in their early teens do not ovulate on every cycle. When we factored the association between age and fecundity and the link between BMI and estimated age into the mathematical model it shifted the optimum fatness down to a BMI somewhere around 17 to 20: exactly where the maximum attractiveness had been in our empirical observations. A series of other studies that have used a whole variety of other image sets and methods have also empirically suggested that the most attractive BMI is generally in the range 17 to 20. Plus remember those pornography models with WHRs all in the 0.65 to 0.7 range, it turns out they also all have BMIs in the range 17 to 21.

The key point is now we know why. On average people rate as the most attractive level of female body fatness the BMI that corresponds to the average BMI of a young female (aged 18 to 22) with maximal future reproductive potential and survival. This ideal transcends ethnicity. Beauty lies in the eye of the beholder, but beholders eyes are the products of evolution and they see signals of reproductive and survival potential, and that is why they almost unanimously agree the level of body fatness in a female that is maximally attractive is around a BMI of 17 to 20.

Who is the fattest of them all, and why?

September 2015

If I was to ask you what the fattest nation on earth is, there is a good chance you would say the United States of America. We are so exposed to media stories that America has a major obesity problem that it might seem obvious that America leads the world in the obesity stakes. But you would be wrong, by a considerable stretch, because in the world obesity league the USA only just manages to scrape into the top ten (at 9th in the world). Before I give you the answer as to which nation is the fattest, it is perhaps worth just thinking for a while about how a country goes about getting this dubious honour. Accurately measuring the amount of fat in a person's body is a difficult thing to achieve. And while there are several different approaches that can do it, like magnetic resonance imaging, computed tomography or dual energy x-ray absorptiometry, all of these methods are expensive, can generally only be performed at specialist facilities, and some of them expose the person being measured to significant amounts of radiation. For making population surveys, that would provide an adequate sample of individuals to characterise the obesity of an entire nation, they are completely unsuitable. To measure how fat people are, at the population level, clinicians and public health bodies need a quick, cheap and easy measurement. The most obvious of which is body weight. Body weight is closely related to the amount of fat that someone carries. There is a problem, however, because body weight not only varies with how fat someone is, but it also depends on how tall someone is as well. Taller people are heavier, even if they are not fatter. Unfortunately, nations not only differ in how fat they are, but also in how tall they are. In the Netherlands in Europe, the tallest nation on earth, for example, the average male height is 1.83m, significantly taller than Americans (at 1.79m) and Chinese at 1.67m, but absolutely towering above the various groups of pygmies from around the world that by definition have mean heights less than 1.5m. If we were to simply measure how heavy people in a nation were, then the Dutch people living in the Netherlands would probably come out near the top, and the pygmies near the bottom. We obviously need some way to account for these height differences.

In theory, one would anticipate that as the linear dimensions of an object (like height) changed then things related to the volume of the same object (like weight) would change in relation to the linear dimension cubed. For example, if I gave you a square cube that measured 10cm along one side. It would have a volume of 10cm x 10cm x 10cm = 1000 cm^3 which is 10^3 cm^3. If I doubled the size of the linear dimension to 20cm, ie increased it by a factor of 2) the volume wouldn't double, but it would equal 20cm x 20cm x 20cm = 8000 cm^3 ie a factor of 8 greater (which is 2^3). On this basis it was felt that the best way to account for the effect of height on weight in humans would be to divide the weight by the height cubed. It turned out, however, that this simple geometric argument was wrong. As long ago as 1848 a Belgian scientist, called Quetelet, measured the heights and weights of a large sample of people, and found that instead of weight being related to the height cubed, it was instead related to the height squared. The reason for this is that as humans get bigger they change shape. Taller individuals are not just short individuals writ larger. For example, our heads stay pretty much the same size independent of how tall we are – with the result that dwarves, for example, end up having what appear to be enormous heads, for the sizes of their bodies. On the other hand as individuals get taller their legs become disproportionately longer. This is particularly noticeable in fashion models. The Chinese model Kong Yansong for example is 175cm tall, which is about 20cm taller than the average Chinese female, but her legs are an amazing 117cm long! The result of this stretching of the limbs, but maintenance of an almost constant head size as height increases, is that weight, which is mostly located in the trunk and head, doesn't increase as fast as we might predict from the geometrical theory. The result is the finding made by Quetelet that weight increases in relation to height squared instead of height cubed. To correct for height therefore it is necessary to divide body weight measurements by height^2. An index was therefore produced called the Body mass index or BMI which equals weight/height^2 and this has been adopted by the World health organisation as the standard measure for body fatness.

The big benefit of BMI is that to measure it you only need a set of scales and a tape measure. It is very cheap, and very quick to measure. There is no radiation exposure for the person being measured, and field workers can go to measure it in large numbers of people, after only minimal training. Once the basic measure was agreed then it was simply a matter of subdividing the scale into categories that correspond to being normal, overweight or obese. This has been based on the relationship between BMI and health parameters. On this basis the 'normal' or healthy range has been set at 18.5 to 25, overweight as 25 to 30 and obese at a BMI of 30+. This division, however, has been the source of some controversy, in part because the relationship between BMI and health is not the same in all populations. In Asians, for example, all cause mortality increases more steeply with BMI than it does in Caucasians. This led to the BMI cut-off points for overweight and obesity in Asians being redefined as 23-28 and 28+ respectively. Another downside of the BMI measure is that at the individual level it can be thrown out by unusual individuals with exceptional body compositions. For example, elite athletes often have very low body fatness but large amounts of muscle for their height. The basketball player Yao Ming, for example, weighed about 138 kg at the peak of his career in the USA, giving him a BMI of 26 and hence well inside the overweight category for an Asian, but in fact his body fatness measured by more sophisticated measurement techniques was only 6%. The anomaly is even more pronounced in body builders. Arnold Swartzenegger, star of the Terminator films and governor of California, in his early career as a body builder weighed 106 kg, and with a height of 188cm would have had a BMI of 30.2 and would hence be classed as obese (other sources suggest his BMI was higher, at 33, well inside the obesity cut-off). Looking at pictures of him from this period it is clear that there is barely a gram of fat on him. However, these are unusual exceptions and at a population level, where elite athletes and body builders are rare individuals, BMI does give a valid snapshot of how fat a nation is.

On this basis the fattest nation on earth in 2007 was the small pacific island of Nauru where 94.5% of people were either overweight or obese (source Forbes : http://www.forbes.com/2007/02/07/worlds-fattest-countries-for beslife-cx_ls_0208worldfat.html). What's more the top seven nations when it comes to obesity were all small pacific islands:

Federated states of Micronesia (91.1%), Cook islands (90.9%), Tonga (90.8%), Niue (81.7%), Samoa (80.4%) and Palau (78.4%). The USA by comparison had 74.1% overweight and obese. In the same listing China was 148th in the world, with only 23.8% of individuals overweight or obese.

An interesting question then is why? What is it about these small pacific islands that make them so susceptible to obesity risk? An idea that has been popular for some time is that this modern day problem is all to do with how the islands were originally colonised, and the conditions when they were first settled by humans. The idea was first published in the 1970s but was heavily promoted, for example, by the popular science author Jared Diamond in more recent years. Colonisation of the Pacific islands seems to have been a gradual spread from the SE Asian land mass which was continuous through Indonesia to the Philippines and New Guinea until at least 30,000 years ago. Island groups such as Fiji and Tonga were settled around 4000 BC and the expansion proceeded until the more remote islands (like Tahiti, Kiribati and Tuvalu) were settled sometime between 500 BC and 1000 AD. Early Polynesian explorers would have had to sail for many days to travel to these remote islands. The classical explanation therefore for why modern day inhabitants of these islands are so fat, is because their ancestors were the individuals who survived such arduous trips of discovery. It is suggested that the only individuals able to survive such journeys were those who carried large amounts of body fat. This would be further selected after arrival because when they arrived they may have had to endure harsh conditions, when obesity would again provide a selective advantage. Hence the islanders would have been originally settled only by individuals carrying genes that predispose to being fat – and their children now carry this legacy.

It is an interesting idea that is appealing because it seems plausible, particularly to a reader who knows nothing about what actually happened, and it is instantly understandable. Plus like all good stories it is based on more than a grain of truth. Fat people do indeed have a survival advantage when it comes to a situation where there is no food to eat. Why this is the case is not immediately obvious. Sure, fat people store more fat reserves, but they also have much higher metabolic rates (and eat substantially more food) and so when there is no food around they also burn through their greater reserves much faster than a lean person does. It turns out, however, from direct measurements of energy demands, combined with the estimated size of the fat stores, that when someone gets fat the increase in stored energy (as fat) is much greater than the increase in energy demands – so they may burn through the energy faster, but not so fast as to destroy the advantage of the fatness. Plus we can be pretty sure this is true, because occasionally people for political reasons starve themselves to death: and it turns out that how long they survive is greater the fatter they are at the start of their fasts.

However, a paper in Annals of Human Genetics this month (September) suggests that this aspect is probably the only part of the classical explanation for why the Pacific islanders are obese that has any truth in it. The paper by Anna Gosling and colleagues, from the University of Otago, in New Zealand, skilfully pulls apart the basis on which the classical model is founded, using archaeological and anthropological data sources from the Pacific

71

region. Gosling and colleagues show that far from being perilous journeys into the unknown, where a group of individuals set off in a canoe hoping to hit land somewhere – and by the time they arrived only the fat ones were still alive, the evidence for excursions of discovery suggests that in fact colonisation was a less random, more planned and considerably more benign process. This involved initial very small groups of scouts making increasingly longer two-way trips of exploration, before larger colonisation parties set out, with known destinations and resources necessary to cover the entire one way trip. Archaeological evidence also suggests that populations on the colonised islands were in regular contact with other island groups, indicating that voyages between them were not perilous one off events, where most people died, but regular low risk occurrences where almost everyone always survived.

So it seems that the colonisation events themselves were unlikely to select for obesity genes. What about selection happening after the colonists arrived: due to the harsh island conditions? Here also the direct evidence for harsh conditions is incredibly weak. Moreover, there is a fundamental mismatch between the rates of mortality that would be necessary to select for obesity genes, and the demographics of the island populations. It seems that it would be impossible to sustain a viable island population and simultaneously select for obesity genes. Finally, perhaps the most damning evidence against the classical idea is that now we are starting to look at the genomes of Pacific islanders nobody has yet found a gene that causes obesity in these populations that is also under positive selection.

In 1902, the British author Rudyard Kipling, who won the Nobel prize for literature 5 years later, wrote a series of short stories, called the 'Just-so' stories. These stories explained for children some of the wonders of natural history – how the camel got his hump, how the leopard got his spots etc. These were all fantastical stories that, however plausible, were based on absolutely no evidence. Gosling and colleagues speculate in their paper that the classic explanation of why Pacific islanders get fat may be another 'Just-so' story: 'How the Pacific islander got his fat'!!

Juice bar

October 2015

Near to my apartment in Beijing a juice bar opened up about a year ago. It is bright green and orange, very plastic and modern, has a few tall uncomfortable plastic seats around high tables, and serves all sorts of interesting concoctions and beverages designed to revive you after the busy working day and generally improve your life. They have intriguing names like, 'mint cleanse', and 'herbal revival' and my favourite 'the antioxidant blitz'. The latter consists mostly of blended blueberries and raspberries, mixed with pomegranate, orange and grape juice. I had one last week, and it was very tasty, almost good enough to take your mind off the plastic furniture you are sitting on. But more than that, it seems that the antioxidant blitz is also a life saver, because if the blurb in the juice bar menu is to be believed, the drink is effective at warding off cancer because it is packed full of antioxidants like vitamin C from the orange, carotenes and carotenoids from the berries, punicalagin (I had never heard of it either!) from the pomegranate, resveratrol from the grape juice, and other unnamed 'antioxidants' which invest it with an unrivalled ability to destroy any nasty free-radicals that may be lurking in your body intent on doing you harm by damaging your DNA, which might lead to development of tumours. It may even be good for neutralising the effects of air pollution: which is always a good potential selling point in Beijing. The whole thing might be convincing if the girl who serves you these miracle cures didn't look half dead from working 15 hours a day, every day, serving juice. If anyone in the shop needs a 'herbal revival' it is her!

I don't know exactly where the juice bar owners got their information from, but if you search on the internet then there is no end of similar material about the life enhancing virtues of increasing your intake of antioxidants. Antioxidants it seems, in addition to keeping cancer at bay, can ward off infections, particularly viral infections like influenza, can alleviate the pain of arthritis, reduce the risk of various neurological disorders such as schizophrenia, autism and depression, improve our memories and reduce the risk of plaque build up in arteries, leading to reductions in heart disease. Antioxidants it seems may even ward off the process of ageing itself. In his book *The Antioxidants*, Richard Passwater, suggests that humans have one of the longest natural lifespans in the animal kingdom, most likely because of the wealth of antioxidants in our omnivorous diet. Human bodies, he also suggested, produce antioxidant enzymes that cannot be found in other creatures (although the details of what these are is lacking). Moreover, the champion anti-oxidant, resveratrol it seems has been found to be so effective at warding off aging-related diseases that it has been dubbed the "fountain of youth." I quite like the antioxidant blitz, but I may have to give it up soon, and the juice bar owner may have to rewrite her menus if recent work published last week in *Science Translational Medicine* is to be believed.

Before I get to that however it is perhaps worth thinking a little about where the whole antioxidant hype comes from. The answer is that it all started in work published over 100 years ago by the German scientist Max Rubner. Rubner worked on measuring the metabolic rates of different animals. In the early 1900s he made a remarkable observation. Larger animals he noted have lower metabolic rates (if you calculate them per gram of tissue). A mouse weighing 25g for example burns through about 75 kJ each day, equal to 3 kJ/g/day. A human however weighing 80000g utilises only 10,000 kJ/day, making its energy use only 0.125 kJ/g/day. Humans, however, also live longer than mice. On average a mouse lives about 3 years and a human about 75 years. Do you see anything interesting about these numbers? The really interesting thing Rubner spotted is that if you multiply the lifespan by the energy expenditure then the result is almost constant. That is 3 x 3 = 9 kJ/gram/life for the mouse and 75 x 0.125 = 9.375 kJ/g/life for the human. Across all the (relatively few) species he looked at the numbers came out the same. Leading to the idea that the amount of energy a gram of tissue uses in a given life is constant, and that animals can choose to burn through energy fast and hence die sooner than if they were more frugal in their energy use. This became known as the 'rate of living theory' of ageing, and it is one of the oldest theories in the field of gerontology.

The main problem with the rate of living theory however was that it was an interesting observation without a mechanism. One really daft idea was that this association is all related to heart beats. That is to supply the high tissue energy demands small animals have greater heart rates, but larger animals with lower energy demands can beat their hearts more slowly – so maybe we are 'given' a fixed number of heart beats in our lives and we slowly work through them until we run out. The magic number of given heart beats it turns out is 2,759 million. When I turned 50 my health insurance company sent me a birthday card saying "congratulations you made your 1,800 millionth heart beat this year". But it didn't make me feel great – just conscious I only had about 1000 million left! As an observation it is roughly true, but as a 'mechanism' it is rubbish because it doesn't explain why we are only given a fixed number or indeed who does the giving.

It wasn't until the 1950s that someone hit on a much brighter idea. In 1954 an Argentinian scientist called Rebeca Gerschman and colleagues at the University of Rochester in New York formulated the idea, published in the journal *Science*, that damage induced by x-rays has the same mechanism as damage from high levels of oxygen: namely both result in the formation of damaging free-radicals. Gerschman noted that free-radicals are probably a by product of normal oxidative metabolism in mitochondria, and that increased metabolism might lead to increased production of free-radicals. Two years later an American scientist Denham Harman made the formal suggestion that these 'free-radical' by-products of metabolism were the primary cause of molecular damage and hence the process of ageing, in a short 2 page article in the Journal of Gerontology. Surprisingly given the prominent location of the published work by Gerschman, it was not cited by Harman. Throughout the 1960s Harmans short note was cited between 1 and 7 times each year. Gerschman'spaper in contrast was cited about twice as often at between 4 and 11 times per year. But then in the early 1970s something happened. Harman published another paper drawing attention to his earlier note. And the citations started to take off.

This was in part because of a fortunate coincidence. If the

free-radical idea is correct then it predicts that a potential way to intervene in the ageing process is to take large doses of antioxidants in the diet. At the time Harman was drawing attention to 'his' idea, Linus Pauling the double Nobel prize winner was extolling the virtues of dosing oneself with enormous doses of vitamin C, as a mechanism to ward off infection but also to combat general ageing because of free-radical damage. Indeed in Harman's paper he says that taking compounds that could neutralise free-radicals (his example was cysteine) would likely have a beneficial effect on the rate of ageing. Harman's short paper from 1956   went from strength to strength and it is now cited almost 300 times per year and has a total citation count of 4000 times. In contrast Gershman's contribution has generated only a total of 500 citations and generally gets less than $1/10^{th}$ the number of cites that Harman gets each year.

Whoever really discovered it, the very attractive part about the free-radical theory is that it provides a cogent mechanism for the rate of living theory. As animals increase their metabolic rate the suggestion is that they potentially generate more free radicals and this increased level of free-radical production leads to greater damage and hence ageing. In the 1990s this idea was the dominant theory for why we age and die. Beating ageing was seen at the time as probably only a matter of reducing the production of free-radicals or the damage they produced. Probably the simplest route into this, it was thought, would be to take lots of antioxidants in the diet. This is when people started to promote foods as being really healthy and 'anti-ageing' if they were packed with antioxidants. It is a trend that has hardly abated over the subsequent 20 years. Unfortunately the science supporting the idea has moved on, but the popular notion of antioxidants being good for you has never caught up. This has been caused at least in part by a series of popular books extolling the virtues of taking antioxidants. Some of the authors of these books are evangelical in their belief of the benefits of taking super high doses of various antioxidants. When my group published a paper a couple of years back showing that when voles (a small mammal like a mouse) are fed high doses of vitamin C they actually live shorter lives, one of the authors of a popular book on antioxidants as an anti-ageing cure-all emailed myself, and the journal that published it, requesting that we publish a retraction because of flaws in the experiment that he threatened to expose if the retraction wasn't forthcoming. We didn't retract but no 'exposure' followed.

The paper in *Science Translational Medicine* published last week (October 7th) concerned the effects of antioxidant supplements on melanoma – which is a rare type of skin cancer (less than 2% of skin cancers are melanomas) but it is the most serious form of skin cancer (almost all skin cancer deaths are melanomas). About 55,000 people worldwide died of malignant melanoma in 2012. The paper was written by Martin Bergo and colleagues from the University of Gothenburg in Sweden. The work was performed in mice that had been genetically engineered to make them susceptible to melanoma. It showed that when the mice were dosed with an antioxidant called NAC (N-acetylcysteine) the melanoma cells changed in a way that made them more likely to metastasise: the process by which cancers spread from one tissue to other tissues in the body, and generally a precursor to mortality. Importantly the dose of antioxidant they used was matched on a body weight basis to the doses that humans routinely take when they take antioxidant supplements. The study showed that there was no difference in the rate at which skin tumours themselves developed between the mice that were fed the antioxidant for 80 days and other control mice fed normal lab chow. However, for the mice fed the antioxidant, the rate at which they developed secondary tumours in their lymph nodes was doubled. This suggested that the antioxidant was somehow changing the ability of the melanoma cells to metastasise. This was confirmed when melanoma cells grown in cell culture were treated with NAC or vitamin E (another antioxidant). This treatment increased the rate at which the cells migrated and were able to invade a nearby membrane. This finding comes on top of another study by the same group published last year which showed that antioxidants can stimulate the growth of lung cancers. So antioxidants appear not to fight existing cancers, as is widely suggested to be the case, but actually to help them grow and spread. The mechanisms appear complex but it seems several different processes are affected. These include externally administered antioxidants allowing cancer cells to preserve higher levels of internally synthesised antioxidants which then help the cells survive attack by the bodies defence system which kills unwanted cells using free-radicals. In addition expression of genes linked to cell movement such as RhoA are increased, but tumour suppressing genes (like p53) are inactivated.

So while antioxidants in our diets, and as supplements, might in theory reduce free-radical production and hence reduce DNA damage and thereby stem mutations that lead to cancer – once a cancer exists, the free-radicals scavengers may serve to preserve and protect these cells, and facilitate malignancy. The solution then would seem obvious: take antioxidants if you don't have cancer, but avoid them if you do. The problem is that in the very early stages cancer is almost impossible to detect. Indeed we may all continuously be generating tumour cells that are identified and killed before they progress. So how do you know if you have a latent tumour or not, and hence if it is safe to take the antioxidants or not?

It makes choosing your beverage at the juice bar extremely difficult. The corresponding author of the Science paper, Dr Bergo, has stated that certainly for established melanoma and lung cancer the work from his lab suggests that taking antioxidants is not going to improve things, and could very well make them worse. Down at my local juice bar I don't see the menu changing any time soon to say "try our antioxidant blitz ….it helps your cancer cells survive and spread". However, next time I am down there I may go for the 'herbal revival', just to be on the safe side. Now I wonder how good the science is behind the claims for that one!.

No pain, no gain

November 2015

About 15 years ago I was involved in the start up of a small pharmaceutical company. The company had two basic lead targets, and I was invited to contribute towards the start-up funds, in return for a share of the company stock, because I had been involved in some of the work which had demonstrated that one of these targets seemed to be effective in causing weight loss in small rodents. The other lead target was a painkiller drug that also was having strong and encouraging effects at reducing pain in assays on rodents. The fortunes of the company like many small pharmaceuticals was a rollercoaster ride, but eventually the company was bought out by a much larger pharmaceutical company, who were primarily interested in the potential weight loss drug, but less so in the painkiller, to which we retained the rights. It was eventually decided by the founding company shareholders that the only way we would manage to get a large pharmaceutical company to show any interest in the pain drug would be to fund our own clinical trial. Clinical trials are where promising drugs that have undergone extensive testing on animals, and are believed on that basis to be safe and effective, are tested on human volunteers. The standard procedure is that a set of volunteers are given either the drug, or a pill that to all intents and purposes looks tastes and feels identical, except it contains none of the active drug. This control is called a placebo. It controls for the potential effects that people might experience simply because they are taking part in a drug trial. The aim of the clinical trials then is to evaluate the drug by how well it performs on the measure of interest (in our case pain relief), relative to the placebo 'drug'.

These clinical trials are normally run by contract organisations who then contact a range of different hospitals and research organisations to do the actual work. An essential component of this process is that the recruited participants (both the patients AND the organisations where they are recruited plus the people giving them the drugs) have no knowledge of which patients have been given the drug and which have been given placebo. The data for each patient is sent in to a central repository of data, managed by the contracting organisation, along with the patient identification code, and generally unless something goes very wrong, like some patients showing strong adverse effects, only at the end of the trial are the codes revealed and a comparison made between the people on the drug and those not on the drug. As you might imagine, given the involvement of paid volunteers, hospitals and research organisations and contract organisations they are expensive to perform. Our trial was fairly small by most standards, yet it still cost millions of yuan to conduct. This money had to come from the pockets of the company shareholders. My wife and I had to personally commit to handing over £20,000 of our savings for the trial to go ahead.

A contract company was engaged to do the work and the trial started. We were initially really excited as we waited for the results, but then we waited and waited until eventually we had almost forgotten about it. Then out of the blue one day we got a message to say that the trial was over. There was good news and bad news. First, the good news: our drug seemed to have a really good effect on pain levels, significantly reducing perceived pain levels in everyone who took it. The bad news was that the same thing also happened with the control group that had been treated with the placebo. They also had also shown large reductions in their levels of perceived pain. For us no pain (in the placebo group) literally meant no gain (in a financial return on our investment to run the trial). Several years have now passed since the trial was completed and we are still looking for a buyer among the big pharmaceutical companies who may be interested in buying up the intellectual property of our pain drug. However, with the results of the clinical trial we funded, nobody seems ever likely to touch it. I had given up some time back of ever seeing our £20,000 again. However, an article published last week in the Journal 'Pain' has given me a renewed glimmer of hope that the whole story may not yet be dead.

Our potential painkiller drug it seems is not unusual. Most drugs that seem to have a promising effect on pain relief in rodents, and have a good safety profile which indicates they are unlikely to have deleterious side effects, then fail to live up to their promise when it comes to clinical trials. In fact, over the last ten years, more than 90% of potential drugs for treatment of neuropathic (nervous system) and cancer pain have failed at advanced phases of clinical trials. The fact this happens so regularly has been used by organisations who are opposed to animal experiments to claim that doing animal studies to identify drugs for humans is a cruel waste of time. This is because animals are forced to suffer pain, before they can be given pain relief, but the physiology of rodents is sufficiently different to that of humans that we can never hope to identify good drugs for human use by this process. We might develop a great drug to alleviate chronic pain in the rat, but that really isn't what the primary aim is. That is one interpretation of the data. Another interpretation however is that the problem doesn't lie in the drugs, or the process of animal testing, but rather in the way that clinical trials are designed: in particular the necessity that a given drug significantly exceed the effects observed in the placebo group. On the face of it the necessity to outshine the placebo is perfectly reasonable. But there is another way of looking at this. If a drug doesn't perform better than placebo in a clinical trial, this doesn't mean that it will necessarily fail in routine clinical use. The critical assumption being made in the trial methodology is that placebo and drug effects are additive. So the difference between drug effect and placebo is the critical diagnostic tool to evaluate the drugs potential use. But what if they are not additive. What if the effects substitute for each other. In other words in the drug treated group the perception of 'being in a drug trial' does not add additional pain relief on top of the drug effect, but is subsumed inside the drug effect. This could happen for example if the drug and the placebo worked by activating the same mechanism. A drug might for example switch on the brains endogenous opiates, to alleviate pain, but the placebo effect might do the same thing. Since there is a limit to how much that system can be stimulated, the effects of drug and placebo would no longer add together in their effects. Critically, this doesn't mean that the drug doesn't work: just that it works in the same way as the placebo does.

As long as the placebo effect is small then this doesn't really matter. For example if a given drug reduced the perception of pain by 25% and the placebo effect for the same type of pain was say 3%. Then it it makes little difference if we assume the effects are additive and take the difference between drug and placebo (=22% improvement) or we assume that the placebo effect substitutes for the drug effect, and so just take the effect of the drug ignoring the placebo (=25% improvement). But if the placebo effect is large then the implications become enormous. For example, if in the above scenario the placebo effect was 24%, then by using the additive model the effect of the drug would only be 1%, but by the substitution model it would still be 25% i.e. 25x more effective than the trial had indicated.

In fact data suggests that drug companies are now finding it significantly harder to develop new drugs. This fact is paradoxical because research technology is getting better and better. Our ability to model the structures of drugs relative to their targets by using computers and computational chemistry has changed out of all recognition compared to 30 or 40 years ago. Why as our research technology gets better and better is it becoming harder and harder to develop new drugs? One answer is that the placebo effect in clinical trials may be getting larger. Of course if the additive model for drug and placebo effects was correct then this wouldn't matter because whatever the placebo effect was the drug effect would just be added on top of it. But if the substitution model was correct, and the placebo effect is subsumed inside the drug effect, then what would happen is the placebo effect might increase over time, while the drug effects would not. Hence it would get more difficult for drugs to do better than the placebo treatments.

The paper just published online in '*Pain*' included an extensive analysis of historical trial data for pain drugs, and found that placebo responses have indeed become larger over the last 25 years, while the impact of the drugs themselves has remained almost static. This alone indicates that the additive model on which clinical trials are based is suspect, unless drug manufacturers with their significantly improved technological capabilities are somehow managing with this technology to invent worse and worse performing drugs. Which of course they aren't, because in animal trials the modern drugs still do well: which is why they go on to be tested in humans in clinical trials. This is why animal trials are so good for evaluating drugs because animals don't show the same placebo effects as humans. The change in the response to placebo treatments for pain, over time was discovered by researchers in Canada, led by Jeffrey Mogil, who directs the pain-genetics lab at McGill University in Montreal. Mogil and colleagues examined 84 clinical trials of drugs for the treatment of chronic neuropathic pain that had been published between 1990 and 2013. Based on patients' pain ratings, the effect of drugs being trialled for pain relief stayed about the same over the 23-year period. However, the placebo effect has steadily increased. In 1996, patients in clinical trials reported that drugs relieved their pain by 27% more than placebo. But by 2013, the improvement over placebo was down to only 9%. Interestingly the increase in the placebo effect was caused by the 35 out of the 84 trials that had been conducted in the USA. Among trials conducted elsewhere there was no significant change in placebo responses. Nowadays, simply being in a trial in the USA for pain drugs, and receiving the placebo treatment seems to relieve pain almost as effectively as many promising new drugs.

The reason for the difference is not clear. However, the same effect of an increasing placebo response has also been observed recently in trials for antidepressant and antipsychotic drugs. This suggests that it is something generic about clinical trials and not something particular to pain. For example, one suggestion is that it might be because as clinical trials in the USA have become longer, larger and more expensive to conduct, they may be enhancing participants' expectations of their effectiveness, and thus increasing the placebo effect of being included into the trial. This may be amplified by the fact that the USA is one of the only countries in the world where drug companies are allowed to market their products direct to customers. This may have generally enhanced peoples expectations of the benefits of taking drugs, and this might create a stronger placebo effect.

Whatever the reason and underlying mechanisms for the elevated placebo effect, the most important finding in the paper is the fact these trends undermine the additive effects model on which all clinical trials are based. What this means is that if a drug stimulates the same pain relieving system that the placebo effect taps into, that doesn't mean the drug doesn't work. We may be throwing away good drug targets that have taken many animal experiments to develop and would become perfectly good drugs because of a too simplistic additive model for the manner in which we conduct and interpret clinical trials. One solution might be for trials not to be conducted by comparing a potential future drug against placebo, but rather comparing a potential drug against the currently best available drug on the market. This way placebo effects become irrelevant. And that is the crack now open in the door for our drug that failed in its clinical trial a few years back, because of the strong placebo response. Maybe there is still a chance we didn't completely lose our £20,000. No pain may hopefully lead to considerable gain if we are lucky.

The protein compass

December 2015

Thirty years ago when I was doing my first research project after finishing my PhD I worked for a while on bats. As part of this work we used to occasionally capture bats from the wild and bring them into captivity to make various measures on them. This was a very pleasant occupation because the bats that we were working on lived in the roof spaces of majestic old houses spread out along the valley of the river Dee, which rises in the Grampian mountains and flows to the east to Aberdeen where it enters the North Sea. In fact two rivers flow into the sea in the city of Aberdeen. At the south of the city is the Dee and at the North of the city is the Don. These two rivers run parallel to each other for about 90 kilometres. The Dee is a rather special valley because it was chosen over 160 years ago by the Queen of England to be the site where she would have a summer home. This home, called Balmoral Castle, was built between 1852 and 1856 and remains part of the Royal Family private estate where the Queen comes late each summer for her summer vacation. Surrounding this estate are lots of very grand old houses and in many of the roofs of these houses there are roosts of brown long-eared bats. So during the summer I would drive around these old houses with my project mentor Professor Paul Racey, and we would take tea and the occasional cake with the people who owned the houses (generally rather delightful old ladies) and then go and poke about in the attics of their houses looking for bats.

Long eared bats normally eat moths that they capture on the wing. Providing such food in captivity is difficult so instead we would feed them on mealworm larvae. The only problem is that bats don't naturally recognise the mealworms are prey, probably because the prey they normally eat are flying around. So once we brought the bats back to Aberdeen we had to train them to eat the mealworms, which was generally achieved by a group of enthusiastic undergraduate students manually feeding the bats until they got the hang of it themselves. To individually identify the bats, and hence keep track of who was learning to feed and who wasn't from their body weights, they had a small clip placed on their forearms that contained a unique number. The reason I am telling you all this is to set the scene for something that happened one summer when we were doing this work. One day Paul and I went out to catch some bats, and we went to a house that is situated right next to the Balmoral castle estate, which housed a particularly large colony. The house was about 60km along the river Dee and then anther 10km along a small tributary. We captured a small number of bats and brought them back to the lab and some students started to train them to feed. But then a disaster happened. One of the students left the lid on one of the cages loose and 3 of the bats escaped. This was pretty annoying because we needed a certain number of the bats to conduct the experiment, which meant we would have to go out and catch more. Paul was busy the next day but the day afterwards we went out to the same roost – and guess what, two of the bats that had escaped were already back in their home colony! We could be absolutely certain that these were the same bats, because they had the little wing clips on with the unique identifying numbers. So that meant within 48h of escaping the bats had managed to navigate their way back home. This is all the more remarkable when I tell you that the Institute where we were holding the bats in captivity is actually located right next to the River Don where it flows into Aberdeen. So the bats hadn't just flown out, found the nearest river and then flown along it until they got to a place they were familiar with, because if they had done that they would have flown up the wrong river. Somehow they had managed to navigate from the 'wrong' river onto the correct valley (the river Dee) and then found their way about 70km home. This was startling because Paul had previously done work on how far the bats in Deeside and Donside travel each night and this work had suggested they seldom travel more than about 5km away from their home roosts. Yet, these bats had been taken about 15x further than they ever normally travel, and had almost instantly found their way back to their home roost. Nights in the summer in Aberdeen are short the total time available to the bats for flying was probably less than 10 hours (assuming that they had only commuted at night). So it seems most likely that the bats had flown directly home. In other words this wasn't the consequence of some lucky trial and error. They simply hadn't had time to randomly relocate their home, they must have known where it was, where they were when they escaped, and how to get between the two places.

Twenty years after our observation in Scotland a group led by Martin Wilelski at Princeton University in the USA studied bat homing behaviour in a different species from that we had observed in Scotland. Their study was a much more sophisticated affair than our anecdotal observation and it was aimed at finding out exactly how bats might navigate over long distances. The bat they studied was the big brown bat. They took bats from roosts and deliberately displaced them 20 km north from their home roosts. A group of bats were released to find their way home. The bats were carrying small radios and they were tracked from a small aircraft. Five km after their release these control bats were on average flying due south, directly back home. But then they did something really sneaky. They exposed some other bats to rotated magnetic fields for a period of 90 minutes around sunset. Some of the bats were exposed to a field rotated 90° clockwise and the others to field rotated 90° anticlockwise. When these bats were also tracked from the aircraft 5km after release the directions they were both heading were almost exactly 90° in error from the correct direction home. Some of these disoriented bats were actually able to eventually compensate and still make it home within the single night. But this was probably because they accidentally encountered a place they were familiar with and were then able to correct their initial navigational error. Bats it seems are probably able to detect the earths magnetic field and they can use this ability (called magnetoreception) to orient themselves.

The ability of animals to sense the earths magnetic fields and thereby orient themselves in space has been suggested to be a feature of animal navigation for well over 100 years. Karl von Frisch for example suggested that honey bees are able to sense magnetic fields to orientate themselves between the hive and feeding areas (in addition to using the position of the sun and polarisation of the sky). Perhaps the most famous studies however referred to the remarkable homing abilities of pigeons that were extensively studied by William Keeton based at Cornell

University in the United States. Keeton showed in 1970 that attaching magnets to homing pigeons interfered with their abilities to get home when the sun was not visible due to cloud cover. There are now many species that are known to have the ability to sense the earth's magnetic fields. The mechanism underlying the capability has however remained elusive. A major step forwards in understanding the mechanism was made in 2008 when it was shown in the fruit fly Drosophila that the ability to sense magnetic fields depends on the gene Cryptochrome 1 (Cry--1). This was shown by training Drosophila to navigate a T-maze in which one arm was exposed to a magnetic field but the other was not. Drosophila lacking functional Cry1 however were unable to perform this task indicating their ability to sense the field was impaired. Interestingly the response was also dependent on the wavelength of light to which the flies were exposed. When flies were exposed to light with wavelengths above 420nm (ie excluding the blue end of the spectrum) they also lost the ability to learn in the maze. This work by Reppert and colleagues was published in Nature. This led to the suggestion 2 years later in 2010, also in Nature and by the same group, that Cryptochrome might facilitate magnetoreception in blue light conditions because under this light Cry-1 becomes activated to form a pair of radicals with opposing spins. The interaction of the earths field with these radicals might then influence the duration that Cry-1 is activated. Because the level of activation of Cry-1 is related to retinal light sensitivity it was suggested that Cry-1 effectively allows Drosophila to literally see magnetic fields.

In the last month (November) however a giant step forwards in this field has been made by a collaboration between several groups in China, based at Peking and Tsinghua Universities and the Chinese Academy of Sciences. This work, led by Professor Xie Can at Peking University and published in Nature materials, has shown that in fact Cryptochromes are part of a large protein complex involving another newly identified protein, that they named MagR. Previously called CG8198, MagR contains iron sulphur (Fe-S) clusters and it polymerises to form a chain of molecules in a double helical arrangement somewhat like DNA. The protein was found by searching the genome to locate proteins that could contain Fe-S clusters. It seems that, together with Cry, these proteins form rod like structures that can be oriented by weak magnetic fields. The researchers showed that this complex is present in many organisms that have magnetoreception, including for example in the eyes of pigeons, leading to speculation that the molecular source of magnetoreception abilities has finally been identified. The discovery has sparked intense interest not only because of the resolution of the molecular mechanism underlying how the Earths field might be sensed, but the potential to use the protein to manipulate organisms.

In the last few years there have been several technological advances that allow researchers to manipulate organisms in ways that were previously impossible. A key technology for example is optogenetics. Using this technique organisms are manipulated to express light sensitive ion channels in neurons. By exposing these neurons to light it is possible to manipulate the activity of neurons expressing the light sensitive proteins and hence define their functions. In 2010 it was highlighted in Science as one of the breakthroughs of the decade. The beauty of the method is that it allows millisecond level control of specific neuronal activity literally at the flick of a switch. Its initial applications were in small organisms like nematode worms and Drosophila. In mammals the technique is more difficult to employ because the skull cuts out much of the incident light on the brain. Hence it is necessary to implant optical fibers into the brain to switch the neurons on and off. One suggestion is that the MagR magnetism sensitive protein might somehow be used to overcome these limitations by coupling its directionality to cellular functions. This could then for example allow neurons to be activated by magnetic fields rather than by light: so called 'magnetogenetics'. This would have the clear benefit that activation could be enabled from outside larger animals without the need for invasive surgery. Indeed back in September a paper was published in the Journal *Science Bulletin* which claimed to have achieved exactly this effect, and coined the name magnetogenetics.

The paper in *Science Bulletin* was published by a group led by Zhang Sheng-jia from Tsinghua university who was previously a collaborator with Xie Can. The paper in Science Bulletin uses the same protein complex discovered by Xie, but it is called MAR rather than MagR. The paper claims to show that worms engineered to express MAR in muscle contracted when exposed to a magnetic field, and that worms engineered to express the protein in touch sensitive neurons also showed a recoil action when the magnetic field was switched on, as if they had been touched. This latter paper has caused a degree of scepticism among the science community, because unlike a light sensitive ion channel it is unclear how aligning this rod like molecule to a magnetic field can cause neurons or muscles to be activated. Even worse publication of the paper in *Science Bulletin*, which spent only 2 days in peer review, while the paper by Xie in Nature materials was involved in a protracted review process lasting almost a year, has sparked an unfortunate row between the groups led by Xie and Zhang, and their respective universities over the specificities of prior research agreements. Whatever the resolution of this row is, there is no doubt that the discovery of this protein complex is a major achievement, and its technological promises may herald a new era in neuroscience similar to that ushered in by optogenetics.

ORIGINAL SOURCES AND FURTHER READING

February   **The lottery you hope you will never win**

Tomasetti, C., and Vogelstein, B. (2015) Variation in cancer risk among tissues can be explained by the number of stem cell divisions. *Science* **347**: 78-81.

March   **When in Rome**

Aplin et al. (2015) Experimentally induced innovations lead to persistent culture via conformity in wild birds. *Nature* **518**: 538-541.

April   **Irresistible drugs**

Ling, L.L. et al. (2015) A new antibiotic kills pathogens without detectable resistance. *Nature* **517**: 455-459.

May   **How dogs make us fall in love with them.**

Nagasawa, M. et al. (2015) Oxytocin-gaze positive loop and coevolution of human-dog bonds. *Science 348*: 333-336

June   **Chocolate for weight loss**

Bohannon, J. (2015) I fooled millions into thinking chocolate helps weight loss. Here's how. Online at

July   **Pandas are cool – its official.**

Nie et al. (2015) Exceptionally low daily energy expenditure in the bamboo-eating giant panda. *Science* **349**: 171-174.

August   **Mathematically the most attractive**

Wang, G.L. et al. (2015) The relationship of physical attractiveness to body fatness. *PeerJ* **3**: e1155

September   **Who is the fattest of them all?**

Gosling, A.L. et al (2015) Pacific populations, metabolic disease and 'just-so' stories: a critique of the 'thrifty geneotype' hypothesis in Oceania. *Annals of human genetics* **79**: 470-480.

October   **Juice bar**

Le Gal K., et al. (2015) Antioxidants can increase melanoma metastasis in mice. *Science Translational medicine* **7**:308, 308re8

Sayin, V.I. et al (2014) Antioxidants accelerate lung cancer progression in mice. *Science Translational medicine* **6**:221: 221ra15

November   **No pain, no gain**

Tuttle, A. H. et al. (2015) Increasing Placebo Responses Over Time in U.S. Clinical Trials of Neuropathic Pain. *Pain* (advance online release) doi: 10.1097/j.pain.0000000000000333

Fava, M. (2015). Implications of a Biosignature Study of the Placebo Response in Major Depressive Disorder. *JAMA Psychiatry,* 1-2.

December   **The protein compass**

Holland, R.A. et al. (2006). Bat orientation using Earth's magnetic field. *Nature* **444:** 702

Gegear, R.J. et al (2008) Cryptochrome mediates light dependent magnetosensitivity in *Drosophila*. *Nature* **454**: 1014-1017

Qin S.Y. et al (2015) A magnetic protein biocompass. Online doi: 10.1038/nmat4484

Long, X.Y. et al. (2015) Magnetogenetics: remote non-invasive magnetic activation of neuronal activity with a magnetoreceptor. *Science Bulletin* online DOI 10.1007/s11434-015-0902-0

*John Speakman was a co-author on the papers by Wang et al and Nie et al.